Donato Nicolò

Search for neutrino oscillations in a long baseline experiment at the Chooz nuclear reactors

TESI DI PERFEZIONAMENTO

SCUOLA NORMALE SUPERIORE
1999

ISBN: 978-88-7642-288-1

Tesi di perfezionamento in Fisica sostenuta il 24 settembre 1999

Donato Nicolò
Laboratori I.N.F.N.
Via Livornese, 1291
I-56010 S. PIERO A GRADO (PI)
Italy

Search for neutrino oscillations in a long baseline experiment at the Chooz nuclear reactors

PACS: 14.60.Pq (Neutrino mass and mixing)

Contents

Contents i

List of Figures v

List of Tables xi

Introduction xiii

I Theory and phenomenology 1

1 Massive neutrino physics 3

 1.1 Overview on the Standard Model 4

  1.1.1 The Higgs mechanism 6

  1.1.2 Fermion masses in the Standard Model 7

 1.2 Neutrino masses in gauge theories 8

  1.2.1 Massive Dirac neutrinos 8

  1.2.2 Neutrino mixing 10

  1.2.3 Majorana neutrinos 10

  1.2.4 Mass hierarchy: the "see-saw" mechanism 11

 1.3 Kinematic tests of neutrino masses 13

  1.3.1 ν_e mass 13

  1.3.2 ν_μ mass 14

  1.3.3 ν_τ mass 15

 1.4 Cosmological constraints 15

  1.4.1 Bounds on neutrino masses 16

2 Neutrino oscillation phenomenology 19

 2.1 Neutrino oscillations in vacuum 19

  2.1.1 The two-flavour scheme 21

2.2 Neutrino oscillations in matter 23
2.3 Experimental hints 24
    2.3.1 The atmospheric neutrino anomaly 24
    2.3.2 The solar neutrino deficit 28
    2.3.3 The LSND result 34
2.4 Experiments at accelerators 36
    2.4.1 The present: CHORUS and NOMAD 36
    2.4.2 Long baseline projects at accelerators 38
2.5 Experiments at nuclear reactors 39
    2.5.1 Previous experiments 40
    2.5.2 Towards long baselines: Chooz and Palo Verde 42
    2.5.3 The future: KamLAND 43

II The Chooz experiment   47

3 The neutrino source   49

3.1 Reactors as ν factories 49
3.2 The E.D.F. power plant at Chooz 51
    3.2.1 Description and working 51
    3.2.2 Reactor power monitor 53
    3.2.3 Map of the reactor core 53
    3.2.4 Fuel evolution 54
3.3 The neutrino spectra 56
    3.3.1 The "summation" approach 57
    3.3.2 The "conversion" approach 58
    3.3.3 Systematic uncertainties of the neutrino spectrum . . . 60
    3.3.4 Neutrino spectrum time relaxation and residual neutrino emission 64
3.4 Neutrino detection 64
    3.4.1 The inverse beta-decay reaction 64
    3.4.2 Other possible targets for low energy neutrinos 66
    3.4.3 Detection techniques 67
3.5 Reactor simulation 68

4 The neutrino detector   73

4.1 The site 73
4.2 The detector 73
    4.2.1 Design and working principle 73
    4.2.2 The target 76
    4.2.3 The geode 76
    4.2.4 The main tank 77

4.3   The scintillators   78

   4.3.1   Study of the optical properties   79

   4.3.2   Scintillator aging tests   81

4.4   The photomultipliers tubes   81

   4.4.1   Selection criteria   81

   4.4.2   The test facility   83

4.5   Detector simulation   84

   4.5.1   General description   84

   4.5.2   The neutron transport code   85

III   Detector setup and calibration   87

5   Neutrino trigger and data acquisition   89

5.1   The main trigger   89

   5.1.1   The first-level trigger   89

   5.1.2   The second level trigger   91

   5.1.3   Trigger for calibration runs   93

5.2   Data acquisition   95

   5.2.1   The on-line system   96

   5.2.2   The digitising electronics   97

5.3   Data structure   99

   5.3.1   The "NNADC" packet   100

   5.3.2   The Reactor Power data   100

5.4   The neural-network based trigger   100

   5.4.1   Motivations   100

   5.4.2   Hardware implementation   101

   5.4.3   Network training   102

6   Detector calibration and event reconstruction   105

6.1   PMT gain   105

6.2   Determination of the photoelectron yield   106

6.3   ADC calibration in a single photoelectron regime   108

6.4   Light attenuation in the Gd-loaded scintillator   108

6.5   Electronics amplification balancing   112

6.6   Event reconstruction techniques   113

   6.6.1   The standard minimization algorithm   113

   6.6.2   Performance   116

   6.6.3   Problems in reconstruction   117

IV   Data analysis   125

7 Neutrino signal and background   127
    7.1   The data sample 127
          7.1.1   Power evolution 128
    7.2   Neutrino selection 128
          7.2.1   Preliminary selection and event classification 128
          7.2.2   Final selection 132
          7.2.3   Positron efficiency 134
          7.2.4   Neutron efficiency 137
          7.2.5   Distance cut efficiency 140
          7.2.6   Neutron multiplicity 140
    7.3   The neutrino signal 142
    7.4   The positron spectrum 143
          7.4.1   Measured spectrum 143
          7.4.2   Predicted spectrum 144
    7.5   The background 145
          7.5.1   Correlated background 147
          7.5.2   Accidental background 148
    7.6   Neutrino yield versus power 149
          7.6.1   Neutrino yield for individual reactors 152
          7.6.2   Neutrino yield versus fuel burn-up 153
    7.7   Neutrino direction 155
          7.7.1   Location of the reactors 155
          7.7.2   Implication for SN neutrinos 158

8 Neutrino oscillation tests   161
    8.1   Integral test (analysis A) 162
    8.2   Two-distance test (analysis B) 165
          8.2.1   Sensisivity to low δm^2 values 165
          8.2.2   Ratios of energy spectra 167
          8.2.3   Comparison of the spectra 168
    8.3   Shape test (analysis C) 170
    8.4   Implications of the Chooz results 173

9 Conclusions   177

Acknowledgements   179

Bibliography   181

List of Figures

2.1 Transition probability for $\sin^2 2\theta = 1$ as a function of L/L_{osc}; the dashed line represents the transition probability (2.8), the solid one the transition probability averaged over a Gaussian energy neutrino spectrum with mean value E and $\sigma = E/10$. 22

2.2 (left) Confidence regions for $\nu_\mu \rightarrow \nu_\tau$ oscillations obtained by Kamiokande and Super-Kamiokande; (right) 90% C.L. allowed region by Kamiokande for $\nu_\mu \rightarrow \nu_e$ oscillations. 26

2.3 (left) Comparison of the (data/MC) ratio vs. neutrino baseline for e-like and μ-like events; the dashed line for μ-like events is the theoretical expectations in the case of $\nu_\mu \rightarrow \nu_\tau$ oscillations with $\sin^2 2\theta = 1$ and $\delta m^2 = 2.2 \cdot 10^{-3}\,\mathrm{eV}^2$. (right) Up-down asymmetry vs. momentum for e-like and μ-like events; the hatched region corresponds to the expectations for the case of no oscillations, whereas the dashed line corresponds to the best oscillation hypothesis. 27

2.4 Main nuclear reactions in the pp cycle involved in thermal energy and neutrino production inside the sun; their contribution to the overall neutrino flux is also listed. 29

2.5 Neutrino energy spectrum for the various pp cycle reactions; the sensitivity region for the different types of experiments is also indicated. 30

2.6 Regions allowed (shadowed areas) to $\nu_e \rightarrow \nu_{\mu,\tau}$ oscillations at 99% C.L. for vacuum (left) and MSW (right) transitions. 32

2.7 (left) Allowed regions (shadowed area) at 90% C.L. by LSND to $\overline{\nu}_\mu \rightarrow \overline{\nu}_e$ oscillations superimposed on the exclusion contours of Bugey3, BNL E734, BNL E776, CCFR and KARMEN. (right) Allowed regions for $\nu_\mu \rightarrow \nu_e$ oscillations (π^+ decay in flight analysis) by LSND at 95% (solid line) compared with the 99% C.L. region allowed by the decay at rest analysis (dotted line). 35

2.8 Expected configuration of a typical $\nu_\tau \mathrm{N} \rightarrow \tau^- \mathrm{X}$ CHORUS event in the emulsion and in the scintillation fibre tracker. 37

2.9   Expected sensitivity to neutrino oscillations for future LBL searches
      at accelerators, compared with the Kamiokande and Super-Kamio-
      kande allowed regions for atmospheric neutrinos and the limits
      from reactor and SBL accelerator experiments. Fig.(a): $\nu_\mu \to \nu_x$
      test. Fig.(b): $\nu_\mu \to \nu_e$ test.   40

2.10 Exclusion plot at 90% C.L. for the 72-day run of the Palo Verde
      experiment; the Chooz 1997 results and the Kamiokande allowed
      region are also shown.   44

2.11 Schematic view of the KamLAND detector.   45

3.1   Mass distribution of the ^{235}U fission fragments.   50

3.2   Working scheme of a PWR reactor.   52

3.3   Schematic view of the fuel rods in the core for the first cycle of the
      Chooz reactors. The number of Boron poison rods assembled with
      each fuel element is also indicated.   54

3.4   Power distribution and burn-up values for the fuel elements in an
      octant of the Chooz reactor core at a certain step ($\beta = 1000$ of the
      first cycle). The contribution to power of each element is normal-
      ized to have mean value equal to one.   56

3.5   Contribution to the fission rate of the relevant fissile isotopes dur-
      ing the first cycle of the Chooz reactors.   57

3.6   Comparison of calculated β^- spectra of ^{235}U fission products and
      the experimental result by Schreckenbach et al. [82]. The dashed
      band indicates the total uncertainties (at 90% C.L.) of the latter. .   59

3.7   Neutrino yield per fission of the listed isotopes, as determined by
      converting the measured β spectra [86, 87].   61

3.8   Comparison of Bugey 3 data with three different reactor spectrum
      models. The error bars include only the statistical uncertainties.
      The dashed lines are the quadratic sum of the quoted error of the
      models and the error due to the energy calibration.   62

3.9   Comparison of the combined (ILL+Bugey) reaction cross section
      with the ILL cross section (left) and their relative error (right) as
      a function of the first Chooz reactor cycle burn-up.   63

3.10 Comparison between the neutrino spectrum at the beginning and
      during the first cycle of the Chooz reactors.   70

3.11 Positron spectra at the startup and during the first cycle of the
      Chooz reactors at maximum daily neutrino luminosity.   71

3.12 Cross section per fission as a function of the reactor burn-up. The
      contribution of each fissile isotope is also shown.   72

4.1   Cosmic muon flux compared to the neutrino flux at the different
      underground experimental sites. In the Chooz case the lower neu-
      trino flux is compensated by the reduction of the muon flux. . . . 74

4.2   Layout of the Chooz detector. 75

4.3   Mechanical drawing of the detector; the visible holes on the geode
      are for the PMT housing (from [96]). 77

4.4   Attenuation length vs. wavelength for the Gd-loaded scintillator
      (of the same type of that used at Chooz) at different aging stages
      (left) and scintillation light attenuation vs. path (right). 80

4.5   Acceleration of the scintillator aging rate as a function of the tem-
      perature. 82

5.1   Number of detected photoelectrons (left) and number of hit PMT's
      (right) for 1 MeV electron events, as a function of the distance from
      the detector centre. 90

5.2   First-level trigger scheme. Both the number of photoelectrons and
      the number of hit PMT's are required to fulfil a certain threshold
      condition. 91

5.3   Electrical scheme of the second-level trigger. 92

5.4   Location of laser flashers and calibration pipes in the detector. . . 94

5.5   Electronics layout for the Chooz experiment, including front-end,
      trigger and digitising modules. 96

5.6   On-line system architecture, with special reference to the different
      bus standards (VME, FastBus, CAMAC) and their interconnection
      (VIC,VMV). 97

5.7   Voltmeter calibration for reactor 1 (left) and 2 (right). 101

5.8   Monte Carlo generated minus CNAPS reconstructed energy (left)
      and z-position (right) for electron events. 103

6.1   Scheme of the electronic chain used to measure the pulse height
      spectrum of sixteen PMT's at a time; each line feeds the signal
      from 8 PMT's. The use of amplified channels is needed for setting
      the gain of the Veto PMT's. 106

6.2   Pulse height spectra for a sample of four PMT's. 107

6.3   Distribution of the single phe peak for all PMT's (left) and its time
      evolution since the start of data taking (right). 108

6.4   Laser efficiency and average number of photoelectrons as a function
      of threshold for a four PMT sample. 109

6.5   ADC spectra obtained with a laser calibration, where the single
      phe peak is clearly visible. The fitting function results from the
      sum of the pedestal and the first three phe's. 110

6.6   Peak associated with the ^{60}Co 2.5 MeV line, as a function of time, as measured by means of a Lecroy QVT. The detected charge follows an exponential decrease, with decay time ≈ 720 d. 111

6.7   Attenuation length measurements at different stages of the data taking period. 112

6.8 λ_{Gd} versus time with best-fit function superimposed. 113

6.9   Comparison of position and energy distributions for runs with the laser flasher at the detector centre, corresponding to three different light intensities. 117

6.10 Data and Monte Carlo distributions of neutron events with the ^{252}Cf source at the detector centre. 118

6.11 Same as before, with the source at $z = -40$ cm. 119

6.12 Same as before, with the source at $z = -80$ cm. 120

6.13 Reconstructed energy spectrum for events associated with neutron capture on Gd. Contributions from γ-lines at 7.94 MeV (capture on ^{157}Gd) and 8.54 MeV (capture on ^{155}Gd) are singled out. The double-Gaussian fit parameters are also shown. 121

6.14 Reconstruction of neutron events with the ^{252}Cf source in region II, $z = 0$. 122

6.15 Energy versus distance from edge for neutron events generated along the calibration pipe. The energy is flat until the reconstructed position is more than 30 cm far from PMT's. 123

6.16 Distributions of neutron events with the ^{252}Cf source at $z = -120$ cm. The discontinuity in the z distribution at the vessel surface is visible also in Monte Carlo generated events. 124

7.1   Power (top) and burn-up (bottom) evolution for Chooz reactors. Both have been off since February 1998. 129

7.2   Neutron versus positron energy for neutrino-like events collected during the reactor-on period. A preliminary cut to the neutron QSUM is applied to reject most of the radioactivity background. . 130

7.3   Same as before, with neutrino-like events collected during the shutdown of both reactors. 131

7.4   Distribution of positron-neutron delay for the different event categories. The best fit curves are also drawn and the relative parameters indicated. 132

7.5   Neutron versus positron energy for neutrino-like events selected from the preliminary sample by applying the "topological" cuts here indicated. 133

7.6   Neutron energy distribution for correlated background events (left) and average E_{n} vs. run number (right). 133

7.7   Determination of the L1lo equivalent energy threshold. The top
       figures show the QSUM spectra measured with a ^{60}Co at the de-
       tector centre by means of an internal and external triggered qVt
       (the corresponding background is superimposed). The central plot
       shows the background subtracted spectra. The bottom histogram,
       displaying the efficiency curve, follows an integral Gaussian func-
       tion whose parameters are indicated. 136

7.8   Equivalent energy threshold at detector centre as a function of
       time. The jumps visible here are due to the threshold setting. . . . 137

7.9   Energy threshold as a function of z for different values of the at-
       tenuation length for the Gd-doped scintillator. The measurements
       obtained with the ^{60}Co source follow the expected behaviour. . . . 138

7.10   Neutron delay distribution measured with the Am/Be source at
       the detector centre (left) and at the bottom edge of the acrylic
       vessel (right). 140

7.11   Neutron energy spectra for reactor-on and reactor-off periods (left)
       and background subtracted spectrum compared to Monte Carlo
       expectations (right). 142

7.12   Same as before, for the positron-neutron distance. 143

7.13   Neutron delay distributions for reactor-on and reactor-off periods
       (left) and background subtracted spectrum compared to MC pre-
       dictions (right). 143

7.14   Experimental positron spectra for reactor-on and reactor-off peri-
       ods after application of all selection criteria. The errors shown are
       statistical. 144

7.15   (above) Expected positron spectrum for the case of no oscillations,
       superimposed on the measured positron spectrum obtained from
       the subtraction of reactor-on and reactor-off spectra; (below) mea-
       sured over expected ratio. The errors shown are statistical. 146

7.16   Energy distribution of e^{+}-like signals associated with the correlated
       background. 148

7.17   Positron yields for the two reactors, as compared with expected
       yield for no oscillations. 153

7.18   Variation of the measured neutrino counting rate, as a function of
       the fuel burn-up for separate (left) and combined (right) reactor
       data and comparison with predictions. Error bars include only
       statistical uncertainties. 154

7.19   Neutron emission angle (with respect to the incident $\bar{\nu}_{e}$ direction)
       vs. its kinetic energy; the discrete structure of lower-left part of
       the picture is an effect of the logarithmic scale for the abscissa
       combined with the $\bar{\nu}_{e}$ energy binning. 157

7.20 Distribution of the projection of the positron-neutron unit vector
      along the incident direction for the selected neutrino sample; the
      data are compared with a higher statistic (5000 events) Monte
      Carlo distribution. 158

8.1   Positron yields for reactor 1 and 2; the solid curves represent the
      predicted positron yields corresponding to the best-fit parameters,
      the dashed one to the predicted yield for the case of no oscillations. 164
8.2   Exclusion plot for the oscillation parameters based on the absolute
      comparison of measured vs. expected positron yields. 166
8.3   Measured ratio of experimental positron yield, compared with the
      predicted ratio in the best oscillation hypothesis (solid line) and in
      the case of no oscillations (dashed line). 168
8.4   Exclusion plot contours at 90% C.L. and 95% C.L. obtained from
      the ratios of the positron yields induced from the two reactors. . . 169
8.5   Exclusion plots at 90% C.L. for the two-distance tests, by using
      the comparison (solid line) and the ratio (dashed line) of the two
      reactor spectra. 171
8.6   Exclusion plot contours at 90% C.L. obtained by the three analyses
      presented. 172
8.7   Results of a three flavour mixing analysis of separate and combined
      Super-Kamiokande and Chooz data, for five representative values
      of δm_{23}. The analysis concerns the 33 kTy data sample for Super-
      Kamiokande and the first Chooz result. 175

List of Tables

2.1   Order of magnitude estimates of the mass sensitivity reachable by different types of neutrino oscillation experiments, both short-baseline (SBL) and long-baseline (LBL). The energies and distances can vary in a wide range and only representative values are listed. 21

2.2   Summary of the results obtained by the solar neutrino experiments; each measurement is compared with the predictions based on the standard solar model [38]. The flux is expressed in SNU units for the radiochemical experiments and in terms of the ^{8}B flux (in units of $10^{-6}\,cm^{-2}\,s^{-1}$) for the water Čerenkov experiments. 30

2.3   Parameters and results of the experiments carried out at reactors. Among these, Palo Verde is the only one currently taking data; Kamland is expected to start operating in January 2001. The δm^2 sensitivity limit, wherever omitted, is obtained at 90% CL. 41

3.1   Energy release per fission of the main fissile isotopes (from ref. [75]). 55

3.2   Time evolution of neutrino spectra relative to infinite irradiation time (from [90]). 64

3.3   Reactor neutrino induced reactions 66

3.4   List of thermal neutron capture reactions 67

4.1   Abundances and thermal neutron cross sections for the Gd isotopes. 75

4.2   Main properties of the liquid scintillators used in the experiment. . 78

7.1   Summary of the Chooz data acquisition cycle from April 7th 1997 till July 20th 98. 128

7.2   Summary of the neutrino detection efficiencies. 142

7.4   Summary of the likelihood fit parameters for the three data taking periods. 151

7.5   Experimental positron yields for both reactors (X_1 and X_2)and expected spectrum ($\tilde{X}$) for no oscillation. The errors (68% C.L.) and the covariance matrix off-diagonal elements are also listed. . . 152

7.6   Measurement of neutrino direction: data and Monte Carlo. 158

7.7   Determination of the Supernova neutrino direction, as obtained
      from Monte Carlo events. The results in the second row are ob-
      tained by requiring the positron–neutron distance to be larger than
      20 cm. Note that the ϕ angle determination is irrelevant since neu-
      trinos are directed along the zenith axis. 159

8.1   Contributions to the overall systematic uncertainty on the absolute
      normalization factor. 163

Introduction

The Letter to the German Physical Society by Pauli in 1930 is considered as the birth of the neutrino physics. The hypothesis of the existence of such a particle was an extreme attempt to explain the continuous energy spectrum of the electrons in β-decays. The Pauli proposal was accepted only because it did not seem more revolutionary than the alternative proposal based on the violation of energy and angular momentum conservation laws.

Even today, seventy years after Pauli's idea and 33 years after its discovery by Reines and Cowan [1], we do not have much more pieces of information about neutrinos than those contained in the quoted letter. Since they interact so weakly with matter, most of their basic properties are still largely unknown. One of the most important issues still to be settled concerns their rest mass. We have no idea why neutrinos are so much lighter than their charged lepton partners; no fundamental symmetry in nature requires massless neutrinos. In the Standard Model this question is solved by assuming *a priori* that neutrinos have no mass, which is consistent with the current results of kinematical tests of neutrino masses. However, this assumption is unsatisfactory from a theoretical point of view, so that many extensions of the Standard Model have been proposed both to give rise to neutrino masses and to explain their smallness in comparison to their charged partners. As a consequence, a clear proof of finite neutrino masses would be a first indication of the existence of new physics beyond the Standard Model.

Moreover, massive neutrinos are demanded to explain the anomalous counting rate of experiments measuring the solar and the atmospheric neutrino fluxes. As a matter of fact, the discrepancy between experimental data and theoretical predictions can be accounted for in terms of neutrino oscillations, which would take place only in the case of massive neutrinos. New strong evidence in favour of neutrino oscillations comes from the latest results by Super-Kamiokande, which confirmed with improved statistics the first indications from IMB, Kamiokande and Soudan [2, 3, 4], about the atmospheric neutrino anomaly. However, these results are still affected by large systematic uncertainties due to the knowledge of the neutrino fluxes. Other experiments utilizing artificial terrestrial sources are then needed to understand the nature of neutrino masses and mixing which are intimately connected with neutrino oscillations.

The subject of this thesis is the search for $\bar{\nu}_e \rightarrow \bar{\nu}_x$ oscillations in CHOOZ, the first long baseline experiment to explore a neutrino mass ($\delta m^2 \gtrsim 10^{-3}$ eV^2) region where hints at neutrino oscillations came from the atmospheric neutrino anomaly [5]. This search was operated at a distance from the reactors (at the nuclear power station at the homonymous village in the Ardennes region of France) of ≈ 1 km, one order of magnitude longer than previous oscillation experiments at reactors [6, 7]. As a result, its sensitivity to small square mass values was improved by one order of magnitude with respect to the limits previously set by any reactor or accelerator experiment.

The neutrino detection was based on the inverse β-decay reaction; the detector was conceived as a homogeneous electromagnetic calorimeter utilizing a Gd-loaded scintillator for neutron identification. Although the neutrino flux at the detector was lower than in previous experiments due to the longer distance, the signal to noise ratio (about 20:1) was preserved thanks to a 115 m thick rock overburden, providing a suppression (by a factor 300 with respect to terrestrial surface) of the cosmic ray flux (which is the most dangerous source of background).

In the first part (including Chapters 1 and 2) we shall deal with the phenomenology of massive neutrinos and neutrino oscillations. The state of the art and the future perspectives in neutrino oscillation searches will be reviewed. In the second part (Chapters 3 and 4) we shall present the experimental layout; a detailed description of the neutrino source (*i.e.* the reactors) and the detector set-up will be given. In Chapter 5 we shall discuss the neutrino trigger concept and describe the on-line system; the data analysis procedures and the event reconstruction technique will be the subject of the following Chapter. The last part will be dedicated to the results; in Chapter 7 we shall discuss the neutrino selection criteria and give an accurate evaluation of the associated efficiencies; in Chapter 8 we shall compare our measurements with expectations to draw our limits on neutrino oscillations.

Part I

Theory and phenomenology

Chapter 1

Massive neutrino physics

The Standard Model assumes massless neutrinos, as we stressed in the Introduction. Up to now this hypothesis is consistent with the results obtained by the terrestrial experiments which provide the following upper limits on neutrino masses:

$$m_{\nu_e} \quad \leq 3.9\,\text{eV} \quad \text{(at 90\% CL)} \quad [8]$$

$$m_{\nu_\mu} \quad \leq 170\,\text{KeV} \quad \text{(at 90\% CL)} \quad [9]$$

$$m_{\nu_\tau} \quad \leq 18.2\,\text{MeV} \quad \text{(at 95\% CL)} \quad [10]$$

Nevertheless, the upper bound are not tight enough to safely exclude a possible existence of massive neutrinos. The ν_μ, for instance, might be about one third as heavy as the electron and the ν_τ might be even 50 times heavier than the electron.

In fact, there is no fundamental reason to assume massless neutrinos in a unified gauge theory of electroweak interactions. In the case of the photon, a zero rest mass is required by the conserved gauge symmetry of the electromagnetic interaction. No such symmetry principle exists in the Standard Model to imply massless neutrinos. Hence this hypothesis is unsatisfactory from a theoretical point of view. Furthermore, indications of massive neutrinos come also from Astrophysics and Cosmology. An experimental search for finite neutrino mass could improve our understanding of essential astrophysical phenomena. As an example, a proof of massive neutrinos could provide a solution to the well-known puzzle of Dark Matter in the Universe and an explanation for the atmospheric anomaly and solar neutrino deficit.

In this chapter we will review the theory of massive neutrinos in the framework of both Standard and non-Standard Models; we will give a cursory look to the theoretical and experimental arguments in favour of massive neutrinos.

1.1   Overview on the Standard Model

The "Standard Model" of electroweak interactions is the most complete theory describing interactions among particles and the most precise in predicting experimental results. From the initial formulation of the electroweak interactions [11], the model has been extended in a more general frame [12] including Quantum Chromodynamics (which describes the phenomenology of hadronic interactions). As far as electroweak interactions are concerned, the model is based on the gauge symmetry associated with the group $SU(2)_L \times U(1)_Y$. This structure was introduced to describe the weak charged current interaction according to the Fermi theory, where only the left-handed components of the fermion fields were involved (from this the suffix L). The representation of a gauge transformation in the $SU(2)_L$ space is written in the form

$$U = e^{ig\alpha(x)\cdot \mathbf{T}} \tag{1.1}$$

where g is the gauge coupling constant of $SU(2)_L$ and $\mathbf{T}$ is the appropriate matrix representation of the group generators. In the case of a doublet, for instance, $\mathbf{T} = \boldsymbol{\sigma}/2$ ($\boldsymbol{\sigma}$ are the Pauli matrices).

The $U(1)_Y$ factor is required by the conservation of the electric charge. The associated gauge generator, the weak hypercharge Y is related to the electric charge and to the third isospin component by the Gell-Mann–Nishima relation

$$Q = T_L^3 + \frac{Y}{2} \tag{1.2}$$

A rotation in the hypercharge space is represented by

$$U = e^{ig'\beta(x)\cdot \frac{Y}{2}} \tag{1.3}$$

g' being the coupling constant relative to the hypercharge.

The theory includes:

- quark and lepton fields assigned to the following representation of $SU(2)_L \times U(1)_Y$:

$$
\begin{aligned}
q_{aL} &= \begin{pmatrix} u_{aL} \\ d_{aL} \end{pmatrix} : \left(2, \tfrac{1}{3}\right) \\
u_{aR} &\qquad\quad : \left(1, \tfrac{4}{3}\right) \\
d_{aR} &\qquad\quad : \left(1, -\tfrac{2}{3}\right) \\
\ell_{aL} &= \begin{pmatrix} \nu_{aL} \\ e_{aL} \end{pmatrix} : (2, -1) \\
\ell_{aR} &\qquad\quad : (1, -2)
\end{aligned}
\tag{1.4}
$$

with $a = 1, 2, 3$ running over the flavour generation:

- gauge boson fields

$$\mathbf{W}^\mu = (W_1^\mu, W_2^\mu, W_3^\mu), B^\mu \tag{1.5}$$

associated with the generators of $SU(2)_L$ and $U(1)_Y$ with $Y = 0$ so that the rule (1.2) holds for all particles.

As stated before, only independent chiral projections $\psi_{R,L} = 1/2(1\pm\gamma_5)\psi$ of the fermions are introduced in order to take into account the $V - A$ structure of the weak charged current interactions. The right handed components are arranged in isosinglet and do not interact with the gauge bosons $\mathbf{W}_\mu$, that is to say, they do not participate in weak charged current interactions. So far the Lagrangian of the model then contains two contributions[13]:

$$\mathcal{L} = \mathcal{L}_G + \mathcal{L}_F \tag{1.6}$$

Here $\mathcal{L}_G$ is a kinetic term of the four gauge bosons and may be written in the form

$$\mathcal{L}_G = -\frac{1}{4}\sum_{a=1}^{3} F_{a\mu\nu}F_a^{\mu\nu} - \frac{1}{2}G_{\mu\nu}G^{\mu\nu} \tag{1.7}$$

where the gauge field tensors are defined as follows:

$$F_a^{\mu\nu} = \partial^\mu W_a^\nu - \partial^\nu W_a^\mu + g\epsilon_{abc}W_b^\mu W_c^\nu, \qquad G^{\mu\nu} = \partial^\mu B^\nu - \partial^\nu B^\mu \tag{1.8}$$

The $\mathcal{L}_F$ term contains the gauge-invariant derivative of the fermion fields:

$$\mathcal{L}_F = \imath\sum_{a=1}^{3}\left[\bar{\ell}_{aL}\gamma_\mu D^\mu \ell_{aL} + \bar{\ell}_{aR}\gamma_\mu D^\mu \ell_{aR} + \right. \tag{1.9}$$

$$\left. + \quad \bar{q}_{aL}\gamma_\mu D^\mu q_{aL} + \bar{u}_{aR}\gamma_\mu D^\mu u_{aR} + \bar{d}_{aR}\gamma_\mu D^\mu d_{aR}\right] \tag{1.10}$$

Here the ordinary space-time derivative ∂^μ is replaced in the kinetic term of the Lagrangian by the covariant derivative

$$D^\mu = \partial^\mu - \imath g\mathbf{T}\cdot\mathbf{W}^\mu - \imath g'\frac{Y}{2}B^\mu \tag{1.11}$$

in order to fulfil the gauge symmetry. Therefore this kinetic term describes both the free motion of the fermion fields as well as their interaction with the gauge bosons. Furthermore the gauge invariance of the theory prevents adding bare mass terms to the Lagrangian. As a matter of fact, a mass term $m\bar{\ell}\ell = m(\bar{\ell}_L\ell_R + \bar{\ell}_R\ell_L)$ would mix left- and right-handed components of the fermion fields, thus violating the required $SU(2)_L$ invariance[14]. At this stage the fermion and boson fields have to be assumed massless.

The extension of the standard model to the strong interactions is obtained by introducing the "colour" quantum number. The theory is then required to be invariant under transformations belonging to the $SU(3)_c$ group. As a result, the local gauge symmetry leads to the introduction in the Lagrangian of four bosons ($W^\pm, Z^0, \gamma$) in the electroweak sector and eight "gluons" (the colour-carrying bosons) in the strong sector.

1.1.1   The Higgs mechanism

We have assumed up to now that both the matter consituents (quarks and leptons) as well as the gauge bosons are massless, which does not agree with the experimental evidence. However, mass terms may arise with the help of the spontaneous breaking of the symmetry according to the Higgs mechanism[15]. In the case of the Standard Model, the spontaneous symmetry breaking is accomplished by introducing a doublet of (complex) scalar fields

$$\phi = \begin{pmatrix} \phi^+ \\ \phi^0 \end{pmatrix} \tag{1.12}$$

The doublet ϕ transforms like ℓ_L under $SU(2)_L$ and has $Y = 1$ in order to maintain (1.2). This contributes to the general Lagrangian by the term

$$\mathcal{L}_H = (D_\mu \phi)^\dagger (D^\mu \phi) - V(\phi) \tag{1.13}$$

where the most general form for the Higgs potential $V(\phi)$ is

$$V(\phi) = -\mu^2 \phi^\dagger \phi + \lambda (\phi^\dagger \phi)^2, \quad \mu^2 > 0 \tag{1.14}$$

The potential (1.14) has infinite, degenerate minimum eigenstates corresponding to a non-zero vacuum expectation value equal to $[\mu^2/2\lambda]^{1/2}$. Let us choose a gauge where

$$\langle \phi \rangle_0 = \begin{pmatrix} 0 \\ \sqrt{\dfrac{\mu^2}{2\lambda}} \end{pmatrix} = \begin{pmatrix} 0 \\ \dfrac{\eta}{\sqrt{2}} \end{pmatrix} \tag{1.15}$$

Notice that the vacuum state is no longer invariant under both $SU(2)_L$ and $U(1)_Y$. In this gauge the kinetic term of the Higgs sector Lagrangian generates the following mass terms for the gauge bosons

$$\mathcal{L}_G^m = \frac{\eta^2}{8}(gW_3^\mu - g'B^\mu)^2 + \frac{\eta^2}{4}g^2 W_\mu^- W^{+\mu} \tag{1.16}$$

where $W^{\pm\mu} = 1/\sqrt{2}(W_1^\mu \pm W_2^\mu)$ are the fields associated with the charged bosons mediating the weak charged current interactions. Let us introduce the mutually orthogonal fields Z^μ, A^μ according to the following

$$Z^\mu = \frac{gW_3^\mu - g'B^\mu}{\sqrt{g^2 + g'^2}} = \cos\theta_W W_3^\mu - \sin\theta_W B^\mu$$

$$A^\mu = \frac{g'W_3^\mu + gB^\mu}{\sqrt{g^2 + g'^2}} = \sin\theta_W W_3^\mu + \cos\theta_W B^\mu \qquad (1.17)$$

θ_W being the Weinberg angle. With the new notation the mass term (1.16) may be rewritten in the form

$$\mathcal{L}_G^m = \frac{\eta^2}{8}\sqrt{g^2 + g'^2}\, Z_\mu Z^\mu + \frac{\eta^2}{4}g^2 W_\mu^- W^{+\mu} \qquad (1.18)$$

from which follows

$$m_W = \frac{g\eta}{2}, \quad m_Z = \frac{\sqrt{g^2 + g'^2}\,\eta}{2}, \quad m_\gamma = 0, \quad \frac{m_W}{m_Z} = \cos\theta_W \qquad (1.19)$$

So the field Z^μ (which is verified to be related to the neutral vector boson) and the charged vector bosons $W^\pm$ acquire a mass, while the field A^μ describing the photon is still massless. This may be stated in another way by observing that the vacuum state of the Lagrangian is invariant under the e.m. gauge transformations. From the relation

$$Q\langle\phi\rangle_0 = \frac{1}{2}(\sigma_3 + Y)\langle\phi\rangle_0 = 0 \qquad (1.20)$$

it follows that the electric charge, which is the generator of the abelian group $U(1)_{EM}$, is conserved. From the Goldstone theorem, a massless boson field A^μ (which may be identified with the photon) is therefore associated with this generator.

1.1.2   Fermion masses in the Standard Model

A possible mechanism to generate fermion masses in the Standard Model consists in the introduction of a Lagrangian term describing the Yukawa coupling of the fermion fields to the Higgs doublet. Such a term may be written in the form

$$\mathcal{L}_Y = \sum_{a,b} \left[f_{ab}^u \bar{q}_{aL}\widehat{\phi}u_{bR} + f_{ab}^d \bar{q}_{aL}\phi d_{bR} + f_{ab}^e \bar{\ell}_{aL}\phi \ell_{bR} \right] + \text{h.c.}, \qquad (1.21)$$

where $\widehat{\phi} = \imath\sigma_2\phi^\star$. After spontaneous symmetry breaking, by using the gauge (1.15), the following mass term

$$\mathcal{L}_m = -\sum_{a,b} \left[\bar{u}_{aL}M_{ab}^u u_{bR} + \bar{d}_{aL}M_{ab}^d d_{bR} + \bar{\ell}_{aL}M_{ab}^l \ell_{bR} \right] + \text{h.c.} \qquad (1.22)$$

is generated. The mass matrices

$$M^i_{ab} = \frac{\eta}{\sqrt{2}} f^i_{ab} \quad \text{with } i = u, d, \ell \tag{1.23}$$

can be diagonalized by means of a biunitary transformation

$$U^{i\dagger}_L M^i U^i_R = D^i \tag{1.24}$$

It is important to note that the neutral and electromagnetic current remains flavour diagonal also in the mass eigenbasis. The charged current instead is no longer diagonal and may be expressed, in terms of the mass eigenstates (u', d', ℓ) as following:

$$J^\mu = \bar{d}'_{aL} \gamma^\mu (1 - \gamma_5) U^{d\dagger}_L U^{(u)}_L u'_L + \bar{e}'_L \gamma^\mu (1 - \gamma_5) U^{(e)\dagger}_L \nu_{eL} \tag{1.25}$$

The quarks partecipating in the weak charged interactions are therefore a combination of the physical states and transitions between quarks of different generations are then possible. The unitary 3×3 mixing matrix $V \equiv U^{(d)\dagger}_L U^{(u)}_L$ is known as the Cabibbo-Kobayashi-Maskawa matrix[16].

The neutrino states are instead massless (and hence degenerate). So the lepton charged current may be diagonalized by redefining the neutrino fields as follows:

$$\nu_{lL} = U^l_L \nu'_{lL} \tag{1.26}$$

The lepton numbers carried by each flavour family is then conserved also by charged current interactions.

1.2   Neutrino masses in gauge theories

1.2.1   Massive Dirac neutrinos

One of the peculiarities of the standard model, as stressed before, is that it contains left and right chiral projections of all fermions except neutrinos. If right-handed neutrinos were introduced in theory (as one does for all the other fermions), the ν_R could pair with the ν_L through the Higgs mechanism to produce a mass term for the neutrinos. Things are arranged so as to predict massless neutrinos in the Standard Model.

Now let us add right-handed Dirac neutrino fields ν_{lR} corresponding to each charged lepton l. Like the other right-handed fields, they are assumed to be $SU(2)_L$ singlets with $Y = 2(Q - T_{3L}) = 0$. Thus

$$\nu_{lR} \quad : \quad (1, 0) \tag{1.27}$$

The presence of these right-handed fields imply new gauge-invariant interactions in the Yukawa sector

$$\mathcal{L}_Y = -\sum_{a,b} f_{ab}^{(\nu)} \bar{\ell}_{aL} \hat{\phi} \nu_{bR} + \text{h.c.}, \qquad (1.28)$$

With a vacuum expectation value as in eq.(1.15), this gives rise to the following mass terms:

$$\mathcal{L}_m^\nu = -\frac{\eta}{2} \sum_{a,b} \bar{\nu}_{aL} f_{ab}^{(\nu)} \nu_{bR} + \text{h.c.} = -\sum_{a,b} \bar{\nu}_{aL} M_{ab}^{(\nu)} \nu_{bR} + \text{h.c.} \qquad (1.29)$$

In general, the mass matrix $M_{ab}^{(\nu)}$ is not diagonal so that ν_{aL}, ν_{aR} do not correspond to the chiral projections of the physical fields. The mass eigenstates may be found by diagonalizing M with a biunitary transformation $U^\dagger M V = m$. Thus, defining new states by the relations

$$\nu_{aL} \equiv \sum_i U_{ai} \nu_{iL}$$

$$\nu_{aR} \equiv \sum_i V_{ai} \nu_{iR} \qquad (1.30)$$

the mass term in eq.(1.29) can be rewritten as

$$\mathcal{L}_m^\nu = -\sum_i \bar{\nu}_{iL} m_i \nu_{iR} + \text{h.c.}, \qquad (1.31)$$

m_i being the ith diagonal element of the matrix m. This equation shows that ν_a are fields with definite masses m_a and are therefore physical particles.

It is thus possible to extend the Standard Model and obtain neutrino mass terms (just like the ones obtained for the other fermions) provided one introduces also right-handed, isosinglet neutrino fields into the theory. However, even this extended model provides no answer to the question about the lightness of neutrinos. A difference in mass between neutrinos and the other particles may be obtained if the coupling constants $f_{ab}^{(\nu)}$ (the ones appearing in the Yukawa term (1.29)) are very small to the corresponding coupling constants which generate quark or lepton masses. But there is no fundamental reason why $f_{ab}^{(\nu)}$ should be small in this model. The inability to predict both the neutrino masses and the magnitude of mixing is one of the shortcomings of the standard model. We will briefly see below that it is possible to put severe restrictions on the neutrino mass spectrum in the frame of GUT models.

1.2.2   Neutrino mixing

One of the major consequences of the introduction of the neutrino mass term
(1.29) in the electroweak Lagrangian is related to the neutrino mixing. To
see this in detail, let us rewrite the charged current interaction between the
W bosons and the leptons in terms of the physical fields ν_i. One obtains

$$\frac{g}{\sqrt{2}}\sum_a \bar{e}_{aL}\gamma^\mu \nu_{aL}W_\mu^- + \text{h.c.} = \frac{g}{\sqrt{2}}\sum_a\sum_i \bar{e}_{aL}\gamma^\mu U_{ai}\nu_{iL}W_\mu^- + \text{h.c.} \qquad (1.32)$$

This shows that in general all neutrinos can have charged current interactions
with a given lepton. In other words, the mass eigenstates are mixtures of the
gauge eigenstates so that a given physical neutrino can couple to more than
one charged lepton via the charged current. This is similar to the mixing
in the quark sector due to the Cabibbo-Kobayashi-Maskawa matrix. As an
immediate result of neutrino mixing, the lepton numbers (L_e, L_μ, L_τ) are no
longer good symmetries. The residual global symmetry in the leptonic sector
is the total lepton number $L_e + L_\mu + L_\tau$.

1.2.3   Majorana neutrinos

Up to now we have treated neutrinos just like the other fermions by assuming
that they are Dirac spinor fields. However, there is an important difference
between the neutrinos and the other fundamental fermions: neutrinos do
not carry any electric charge. This leads to the theoretical possibility that
the neutrinos are Majorana spinors. If this were the case, neutrinos and
antineutrinos would be just the opposite chiral states of the same particle
which are CPT-conjugates of each other.

By using the left- and right-handed fields $\nu_{L,R}$ and their conjugate $\nu_{L,R}^c$,
it is possible to build additional, Lorentz-invariant bilinear mass terms for
Majorana neutrinos such as

$$m_L \bar{\nu}_L \nu_L^c, \quad m_R \bar{\nu}^c_R \nu_R \qquad (1.33)$$

It is easy to see that these mass terms would be forbidden for the other
fundamental fermions because of the charge conservation. As a matter of
fact the fermion fields behave under a global phase transformation as follows:

$$\psi_L \stackrel{U(1)}{\rightarrow} e^{\imath Q\theta}\psi_L, \psi_L^c \stackrel{U(1)}{\rightarrow} e^{-\imath Q\theta}\psi_L^c \Rightarrow \bar{\psi}_L\psi_L^c \stackrel{U(1)}{\rightarrow} e^{-2\imath Q\theta}\psi_L \qquad (1.34)$$

Eq. (1.34) implies that Majorana neutrinos would give rise to transitions
with $\Delta L = \pm 2$ (such as neutrinoless double-beta decays) which violate the
conservation of the lepton number.

By adding the terms (1.34) to eq. (1.29), we may rewrite the mass Lagrangian in the matricial form

$$\mathcal{L}_m^{(\nu)} = \frac{1}{2}(\bar{\nu}_L \ \bar{\nu}_R^c)\mathcal{M}\begin{pmatrix} \nu_L^c \\ \nu_R \end{pmatrix} + \text{h.c.} \tag{1.35}$$

where, in the simplest case of one generation only,

$$\mathcal{M} = \begin{pmatrix} m_L & m_D \\ m_D & m_R \end{pmatrix} \tag{1.36}$$

The mass matrix $\mathcal{M}$ may be diagonalized by means of an orthogonal matrix

$$\mathcal{O} = \begin{pmatrix} \cos\theta & -\sin\theta \\ \sin\theta & \cos\theta \end{pmatrix}, \quad \text{with } \tan 2\theta = \frac{2m_D}{m_R - m_L} \tag{1.37}$$

and its eigenvalues are

$$m_{1,2} = \frac{1}{2}\left[m_L + m_R \pm \sqrt{(m_R - m_L)^2 + 4m_D^2}\right] \tag{1.38}$$

The eigenvalues of the general mass term are not necessarily positive. If $m_2 < 0$, the required positive value of the mass eigenvalue has to be restored by redefining the corresponding eigenvector $\nu_2 \to \gamma_5\nu_2$ (whose effect is a change of sign of the charge conjugation eigenvalue) [17].

Thus, for a single family, we obtain the two eigenstates

$$\begin{aligned} \nu_1 &= \sin\theta(\nu_L + \nu_L^c) + \cos\theta(\nu_R + \nu_R^c) \\ \nu_2 &= \cos\theta(\nu_L \pm \nu_L^c) - \sin\theta(\nu_R^c \pm \nu_R) \end{aligned} \tag{1.39}$$

where the negative sign arises if $m_2 < 0$. One sees immediately that the neutrino mass eigenstates are also charge conjugation eigenstates, that is to say, they describe Majorana particles. For N generations, one obtains $2N$ Majorana particles in general.

If $m_L = m_R = 0$, the mass Lagrangian (1.35) reduces to the usual Dirac term (1.29). Hence Dirac neutrinos may be thought as the superposition of two degenerate Majorana fields ($m_{1,2} = m_D$). So the mass eigenstates (1.39) may represent one Dirac neutrino (as in the extended Standard Model), two Majorana neutrinos, or a Majorana neutrino and a Majorana antineutrino [17].

1.2.4 Mass hierarchy: the "see-saw" mechanism

The most interesting aspect of the model with Majorana neutrinos is that it can provide a reason for the smallness of neutrino masses. To see this, let us refer to the one-generation model and assume that, besides the ν_L, which couples to the lepton in the weak charged current, there exists an isosinglet Majorana neutrino ν_R with mass m_R. In the mass matrix, the quantity m_D arises from the Yukawa coupling and therefore it is natural to assume that it is of the same order of magnitude as the masses of the other fermions in the same family. On the other hand, the right-handed neutrino must be heavy $(m_R \gg m_Z)$ otherwise it would contribute to the number of neutrino species as determined at LEP. In the frame of GUT theories, we may assume a hierarchy between masses such that

$$m_L \approx 0, \quad m_R \approx m_{GUT}, \quad m_D \approx m_f \tag{1.40}$$

where m_{GUT} is the GUT mass scale ($m_{GUT} \approx 10^{16}$ GeV in the minimal SU(5) model). The mass matrix then reduces to

$$\mathcal{M} \simeq \begin{pmatrix} 0 & m_D \\ m_D & m_R \end{pmatrix} \tag{1.41}$$

and its eigenvalues become

$$m_1 \simeq m_R, \quad m_2 \simeq \frac{m_D^2}{m_R} \quad \Rightarrow \quad m_1 m_2 \simeq m_D^2 \tag{1.42}$$

Since $m_R \gg m_D$, it follows that $m_2 \ll m_D$, which means that the left-handed neutrino is much lighter than the charged fermions [1]. This trick to generate neutrino masses, which is better knows as the "see-saw" mechanism, is unique to neutrinos (since it is based on the Majorana mass) and it is realized in a natural way in some extensions of the Standard Model.

   This method may be extended to all the fermion families. If one assumes that m_R has the same value for all three generations, then it is possible to obtain a relation among the masses of the three light neutrinos if one takes for m_D the value of either the charged lepton or the quark with $T_3 = 1/2$ in each generation. In the former case one expects that

$$m_{\nu_e} \div m_{\nu_\mu} \div m_{\nu_\tau} = m_e^2 \div m_\mu^2 \div m_\tau^2$$

while in the latter case $\tag{1.43}$

$$m_{\nu_e} \div m_{\nu_\mu} \div m_{\nu_\tau} = m_u^2 \div m_c^2 \div m_t^2$$

According to this model, the τ neutrino (the heaviest one) could be heavier than the electron neutrino (the lightest one) by about 10 order of magnitudes.

---

[1] In fact the mass eigenstates have no longer a well-defined chirality, since they are linear combinations of left and right chiral projections. But if the (1.40) holds, the mixing angle is very small ($\theta \simeq -m_D/m_R$); hence the rotation matrix (1.37) is very close to the identity and the mass eigenstates are roughly equal to the chirality eigenstates.

Modification of this idea have been proposed, often replacing the m_R by a smaller mass scale so as to increase the mass of light neutrinos. If the mass of right-handed neutrinos follow the family hierarchy, the quadratic dependence of eq.(seemass) is replaced by a linear dependence on the charged fermion masses. It has been pointed out [18] that the use of a "linear" see-saw mechanism in the minimal SUSY $SO(10)$ model can accomodate both the solar as well as the atmospheric neutrino puzzle (which we will deal with in the next Chapter), with $m_{\nu_\mu} \simeq 1.7 \cdot 10^{-3}\,\mathrm{eV}^2$ and $m_{\nu_\tau} \simeq 5.9 \cdot 10^{-2}\,\mathrm{eV}^2$.

1.3 Kinematic tests of neutrino masses

One direct way to determine neutrino masses consists in looking at the kinematics of decay processes involving neutrinos. Let us review the theoretical ideas on which the experiments directly searching for a neutrino mass are based, while discussing the most significant results obtained so far.

1.3.1 ν_e mass

In a nuclear beta decay

$$(A, Z) \rightarrow (A, Z+1) + e^- + \bar{\nu}_e \qquad (1.44)$$

the parent and the recoil nuclei are much heavier than the electron. Thus the outcoming electron and $\bar{\nu}_e$ share the total available energy Q. Although the $\bar{\nu}_e$ is hard to detect, measurements on the escaping electron can provide information about the $\bar{\nu}_e$. The electron kinetic energy spectrum, calculated by using the Fermi interaction, is[17, 19]

$$n(E) = \frac{G_F^2 m_e^5 \cos^2\theta_c}{2\pi^3}|\mathcal{M}|^2 F(Z, E)pE(Q - E)\sqrt{(Q - E)^2 - m_{\nu_e}^2} \qquad (1.45)$$

where $\mathcal{M}$ is the nuclear matrix element and $F(Z, E)$ is the correction factor due to the Coulomb interaction in the final state. If $m_{\nu_e} = 0$, from eq.(1.45) one obtains

$$K(E) = \left[\frac{n(E)}{F(Z, E)pE}\right]^{1/2} \propto Q - E \qquad (1.46)$$

In the case of a finite electron neutrino mass, $K(E)$ is no more linear with the electron energy and vanishes for $E \geq Q - m_{\nu_e}$. Hence the measurement of the shape of the electron spectrum allows us to obtain direct information on the ν_e mass. The effect of this mass becomes appreciable only near the end-point of the spectrum where $Q - E$ is comparable with m_{ν_e}. It is straightforward to obtain that, in absence of a neutrino mass, the function $n(E)$ has a factor $(Q-$

$E)^2$, so that the fraction of decays producing electrons in the energy range from $Q - \Delta E$ to Q is proportional to $(\Delta E/Q)^3$. It is therefore important to choose a β-decay process with a Q value as small as possible.

The most commonly used is the tritium decay, which occurs with the lowest known $Q = 18.6\,\text{KeV}$. Even in this case, the rate of electrons near the end-point is extremely low (10^{-9} for $\Delta E \approx 10\,\text{eV}$). It is therefore necessary to use very intense ^3H sources as well as to minimize the background. The sources must be very thin (of the order of a few atomic layers) in order to minimize the residual path of electrons. Apart from these, there are other problems affecting the analysis of experimental data due to the atomic electrons surrounding the decaying nuclei. First, they can modify the β spectrum shape. Second, the Coulomb correction factor is affected since the atomic electrons screen the nuclear charge.

Most of the high precision experiments performed with tritium actually saw an excess of events near the end-point; the fitted neutrino mass values were always non-physical ($m_\nu^2 < 0$) and the upper limits were poorer than their theoretical sensitivity. However, the last measurement obtained by the Troitsk group [8] is $m_\nu^2 = (1.5 \pm 5.9 \pm 3.6)\,\text{eV}^2$, corresponding to the upper limit

$$m_{\nu_e} < 3.9\,\text{eV (at 90\% C.L.)} \tag{1.47}$$

1.3.2 ν_μ mass

The kinematic search for a muon neutrino mass is accomplished by the analysis of charged pion decay

$$\pi^+ \rightarrow \mu^+ + \nu_\mu \tag{1.48}$$

This decay channel has a branching ratio very close to unity ($1 - \Gamma_i/\Gamma \simeq 1.22 \times 10^{-4}$) and the 2-body final state makes the kinematical analysis very easy. If the decaying pion is at rest, one has:

$$p_\mu^2 = \frac{(m_\pi^2 + m_\mu^2 - m_\nu^2)^2}{4m_\pi^2} - m_\mu^2 \tag{1.49}$$

It is thus possible to determine the ν_μ mass if one knows the muon and the pion masses and measures the muon linear momentum.

Since the pion and muon masses are much larger than the neutrino mass, the muon momentum is rather insensitive to m_ν. As an example, the muon momentum would change by about 3×10^{-5} with a neutrino mass ranging from 0 to 250 KeV. A very accurate measurement of the muon momentum is thus required. Another drawback of this procedure comes from the knowledge of the pion and muon masses. Small uncertainties in these masses may induce a large error on m_ν.

The experiment operated at the muon spectrometer at PSI in Villigen (Switzerland)[9] used this method to set the most stringent limit obtained up to now:

$$m_{\nu_\mu} < 170\,\text{KeV} \qquad \text{at } 90\% \text{ C.L.} \tag{1.50}$$

1.3.3 ν_τ mass

The most sensitive experiments searching for the ν_τ mass make use of the analysis of the τ semileptonic decay modes, which are allowed thanks to its large mass (1.784 GeV). Since a lot of the decay energy goes into the hadron (more frequently pion) masses, the neutrino shares a small amount of that energy, thus increasing the sensitivity to the mass.

This procedure is followed by the experiments working at e^-e^+ colliders (such as LEP, DESY). The $\tau^+\tau^-$ pairs produced from the e^-e^+ collisions are identified by the observation of a simple decay mode of one of the τ's (typically one charged particle and neutrinos). The events where the other τ decays in the multipion mode

$$\tau^\pm \rightarrow 3\pi^\pm 2\pi^\mp \nu_\tau(\overline{\nu}_\tau)$$
$$\text{or} \tag{1.51}$$
$$\tau^\pm \rightarrow 3\pi^\pm 2\pi^\mp \pi^0 \nu_\tau(\overline{\nu}_\tau)$$

are then selected and the missing energy and momentun are reconstructed to determine the ν_τ mass.

The ALEPH experiment[10] performed an analysis on a sample of $\approx$ 180000 $\tau^+\tau^-$ pairs; 52 events were selected with 5 charged pions (7% of which in the six-pion mode). The analysis included 2939 events with τ-decays in the three-pion mode; the resulting upper limit on τ neutrino mass was improved at 18.2 MeV at 95% C.L.

1.4   Cosmological constraints

In the standard big-bang cosmology, neutrinos are the most abundant form of matter in the Universe next to radiation. Their abundance is such that, if neutrinos had masses of several eV and their lifetime were comparable to or longer than the lifetime of the Universe, neutrino mass would dominate the total mass of the Universe. Therefore massive neutrinos would have deep effects on the expansion of the Universe, from its beginnings (primordial nucleosynthesis of light elements) up to the present state. Cosmological considerations provide important restrictions on the neutrino mass and on the number of possible neutrino species.

1.4.1   Bounds on neutrino masses

The standard theory of the evolution of the universe predicts the existence of a primordial "neutrino sea" whose density should be of the same order of magnitude as the blackbody background radiation discovered in 1964 by Penzias and Wilson. Of course the experimental observation of this neutrino cosmic background is an objective much more difficult to achieve and seems to be out of our actual possibilities.

Nevertheless, their concentration can be estimated by assuming that neutrinos, in the early epochs of the universe, were in thermal equilibrium with the photons and the other relativistic particles. In an expanding Universe, the temperature is rapidly decreasing; particles are in equilibrium as long as their energy changes faster than the temperature, i.e. their interaction rate is faster than the expansion rate of the Universe [20]. In the usual Robertson-Walker expanding Universe, the expansion rate is $H \sim T^2/M_P$, where $M_P \sim 10^{19}$ GeV is the Planck mass and T is the Universe temperature. On the other hand the cross section for neutrino interaction is of the order of $G_F^2 T^2$, while the number density of the involved particles is $\sim T^3$; therefore the typical neutrino interaction rate is given by

$$\Gamma_\nu \approx G_F^2 T^5 \qquad (1.52)$$

where $G_F = 10^{-5}$ GeV^{-2} is the Fermi constant. Then neutrinos decoupled from the Universe expansion as soon as the temperature fell below the value

$$T_d \simeq \left(\frac{1}{G_F^2 M_P} \right)^{1/3} \sim 1\,\text{MeV} \qquad (1.53)$$

If neutrinos were relativistic at the time of decoupling ($m_\nu \ll T_d$), their density would be comparable to that of photons; by applying the Fermi and Bose distributions to the neutrino and photon gases, one obtains that

$$n_\nu(T_d) = \frac{3}{4} n_\gamma(T_d) \qquad (1.54)$$

The ratio of neutrino to photon density remained constant until $T \approx 5 \times 10^{9}\,^0 K = 0.5\,\text{MeV}$ ($t \approx 4\,\text{s}$), when electrons and positrons ceased to be in equilibrium with the radiation; the $e^+ e^-$ annihilation increased the photon density by a factor 11/4 (due to the entropy conservation). Hence the present neutrino density is [21, 19, 17]

$$n_\nu = \frac{3}{11} n_\gamma \simeq 110\,\text{cm}^{-3} \qquad (1.55)$$

where we used $n_\gamma = 413\,\text{cm}^{-3}$ (corresponding to a blackbody temperature of $T = (2.73 \pm 0.05)\,^0 K$); the contribution of neutrinos to the present energy density of the Universe is

$$\rho_\nu = n_\nu \sum_i m_{\nu_i} \tag{1.56}$$

The most relevant constraint on neutrino masses follows from demanding that the neutrino energy density does not overclose the Universe. It is likely [22, 23] that the Universe energy density is not far from the critical density

$$\rho_c = \frac{3H_0^2}{8\pi G_N} \simeq 1.9 \times 10^{-29} h^2 \frac{\text{g}}{\text{cm}^3} \simeq 1.05 \times 10^4 \frac{\text{eV}}{\text{cm}^3}, \tag{1.57}$$

$H_0 = 100\,\text{km/s/Mps}$ being the current value of Hubble's constant and $h = 0.65 \pm 0.1$ [23]; the energy density (1.57) is the transition value from "open" Universe (with $\rho < \rho_c$) to "close" Universe ($\rho > \rho_c$). The observation of the luminous parts of galaxies (i.e. baryonic matter) gives a rather small value ($\Omega \sim 0.01$) [23]. Another determination, based on the dynamics of the galactic motion, infers an average density $\Omega \sim 0.2$, a quantity much larger than the density of baryonic matter, but still insufficient for Universe closure (the discrepancy between the two observations leads to the well-known dark matter problem). Relic neutrinos could bridge the gap between the observational values and the critical density; due to their abundance, their contribution to the Universe density will be significant even in the presence of small masses. By defining $\Omega_\nu \equiv \rho_\nu/\rho_c$, the closure condition leads to

$$\Omega_\nu = \frac{\sum_i m_{\nu_i}(\text{eV})}{92h^2} < 1, \tag{1.58}$$

which is satisfied if the sum of neutrino masses does not exceed 35 eV. It is interesting to note that this limit is comparable with the upper limit obtained for the ν_e and is less than the limits on ν_μ, ν_τ by four and six orders of magnitudes respectively; muon and tau neutrino masses must be well below their kinematical bounds.

Chapter 2

Neutrino oscillation phenomenology

The neutrino oscillation hypothesis was first proposed by Pontecorvo [24] in 1957, a few years before the discovery of the other neutrino flavours. The idea was that the neutrino state produced in weak interaction processes is a superimposition of two Majorana neutrinos with definite masses; hence the possibility of neutrino elicity oscillations, analogous to the strangeness oscillations in the neutral K-meson system. Only several years later (Maki *et al.* [25] in 1962, Pontecorvo [26] in 1967) the same formalism was extended to describe possible lepton flavour-changing transitions. The neutrino oscillation hypothesis descends in a natural way from the mixing of the neutrino flavours, which is related to the introduction of neutrino mass terms into the electroweak Lagrangian. The left-handed flavour neutrino fields are unitary linear combinations of the left-handed components of (Dirac or Majorana) massive neutrino fields; the time evolution of the initial field multiplies each component by a phase factor varying with time and hence with the travelled distance. The counting rate of a particular neutrino flavour is thus a periodical function of the source-detector distance.

This phenomenon can take place only in the case of massive neutrinos. An experimental evidence of neutrino oscillations could be an unquestionable proof of the existence of massive neutrinos. Moreover, a measurement of the oscillation amplitude could provide information about the mass values (of their differences, to be precise) and the mixing matrix elements.

2.1   Neutrino oscillations in vacuum

Let us consider a pure ν_l beam, generated at $x = 0$ and at time $t = 0$ by a CC reaction involving the associated charged lepton l. At the production point the neutrino field is described by a coherent superimposition of mass eigenstates ν_i as follows:

$$\nu_l = \sum_i U_{li}^* \nu_i \qquad (2.1)$$

The mass eigenstates evolve in time with the phase factor $\exp(-\imath E_i t)$; the neutrino state at time t is then described by

$$\nu_l(t) = \sum_i U_{li}^* \exp(-\imath E_i t) \nu_i \qquad (2.2)$$

The neutrino detection, at a distance L from the source, projects the neutrino state on the flavour state $\nu_{l'}$ (the one the detector is sensitive to); the detection probability amplitude is then given by

$$\mathcal{A}_{\nu_l \to \nu_{l'}}(t) = \sum_i U_{l'i} \exp(-\imath E_i t) U_{li}^*; \qquad (2.3)$$

using the unitarity relation $\sum_i U_{l'i} U_{li}^* = \delta_{ll'}$, the transition probability can be written in the form

$$P_{\nu_l \to \nu_{l'}} = |\mathcal{A}_{\nu_l \to \nu_{l'}}|^2 = \left| \delta_{ll'} + \sum_{i \neq j} U_{l'i} U_{li}^* \left[\exp\left(-\imath \frac{\delta m_{ij}^2 L}{2p}\right) - 1 \right] \right|^2, \qquad (2.4)$$

where $\delta m_{ij}^2 \equiv m_i^2 - m_j^2$ and $E_i \simeq p + m_i^2/2p$ in the ultrarelativistic approximation (p is the beam momentum).

The transition probability (2.4) contains a periodical dependence on the travelled distance L (hence the term "oscillations"), the period being

$$L_{osc} = 4\pi \frac{p}{|\delta m_{ij}^2|} \simeq 4\pi \frac{E}{|\delta m_{ij}^2|} = 2.48 \frac{E(\text{MeV})}{|\delta m_{ij}^2|(\text{eV}^2)} \qquad (2.5)$$

which is the so-called "oscillation length". Neutrino transition can be observed only if the neutrino mixing takes place and the travelled distance is longer than the oscillation length, which holds if at least two neutrino mass eigenstates satisfy the condition

$$\delta m_{ij}^2 \gtrsim \frac{E}{L} \qquad (2.6)$$

Hence a search for neutrino oscillations can provide information about the squared mass differences; the larger the value of the ratio L/E (which is commonly referred to as "baseline"), the smaller are the values of δm^2 which can be probed. Tab. 2.1 resumes the sensitivity to δm^2 of different types of neutrino oscillation experiments. In most cases, the sensitivity limit is quite beyond the upper bounds set by the kinematic tests. One can conclude that neutrino oscillations is the most powerful technique to explore little mass difference values.

The neutrino oscillation experiments are usually classified into:

Table 2.1: Order of magnitude estimates of the mass sensitivity reachable by different types of neutrino oscillation experiments, both short-baseline (SBL) and long-baseline (LBL). The energies and distances can vary in a wide range and only representative values are listed.

Source	Flavour	$L(\,\mathrm{m})$	$E(\,\mathrm{MeV})$	$\delta m^2(\,\mathrm{eV}^2)$
Sun	ν_e	$1.5 \cdot 10^{11}$	1	10^{-11}
Atmosphere	$\nu_\mu(\overline{\nu}_\mu), \nu_e(\overline{\nu}_e)$	$10^4 \div 10^7$	10^3	10^{-4}
Reactors (SBL)	$\overline{\nu}_e$	10^2	4	10^{-2}
Reactors (LBL)	$\overline{\nu}_e$	10^3	4	10^{-3}
Accelerators (SBL)	$\nu_\mu(\overline{\nu}_\mu)$	10^3	10^3	1
Accelerators (LBL)	$\nu_\mu(\overline{\nu}_\mu)$	10^6	10^3	10^{-3}

- *appearance* experiments;

- *disappearance* experiments.

The objective of the former is to find evidence for neutrinos $\nu_{l'}$ in a (nearly pure) ν_l beam (with $l \neq l'$), which would imply a non-zero transition probability. Experiments of this type are possible only if the neutrino beam energy is above the threshold to produce the associated charged lepton l' by CC interactions. As indicated in Tab. 2.1, only experiments detecting atmospheric or accelerator neutrinos fulfil this condition. An additional requirement is that the $\nu_{l'}$ contamination in the beam be kept as low as possible (otherwise it should be known very accurately) in order to minimize the background.

The "disappearance" experiments usually compare the measured ν_l flux with respect to expectation to find a significant deficit. If the source is not accurately known, an oscillation pattern can be searched for by measuring the ν_l flux at varying distances from the source. These experiments are thus sensitive to the transition $\nu_l \rightarrow \nu_x$, x being any possible lepton flavour.

2.1.1 The two-flavour scheme

The results of neutrino oscillation experiments are usually analysed in the simpler scheme of a two-flavour oscillation. In this frame the mixing matrix (which in the most general case contains three mixing angles plus one CP-violating phase) can be rewritten in terms of one mixing angle θ; for instance, in the simple case of oscillations between electron and muon neutrinos, eq.(1.30) becomes

$$\begin{pmatrix} \nu_e \\ \nu_\mu \end{pmatrix} = \begin{pmatrix} \cos\theta & \sin\theta \\ -\sin\theta & \cos\theta \end{pmatrix} \begin{pmatrix} \nu_1 \\ \nu_2 \end{pmatrix} \tag{2.7}$$

The transition probability reduces to

$$
P_{\nu_l \to \nu_{l'}} =
\begin{cases}
1 - \sin^2 2\theta \sin^2 \left(\dfrac{1.27 \delta m^2 (\mathrm{eV}^2) L(\mathrm{m})}{E(\mathrm{MeV})} \right) & \text{if } l = l', \\[2ex]
\sin^2 2\theta \sin^2 \left(\dfrac{1.27 \delta m^2 (\mathrm{eV}^2) L(\mathrm{m})}{E(\mathrm{MeV})} \right) & \text{if } l \neq l'.
\end{cases}
\tag{2.8}
$$

However, this result applies to the ideal situation where both the source and
the detector are pointlike and the beam is monoenergetic. In fact one measures a probability averaged over the finite size of both source and detector
and over the width of the energy distribution; as shown in Fig. 2.1, this averaging produces a damping of the oscillation amplitude, which practically
reduces the transition probability to be constant $(1/2 \sin^2 2\theta)$ for $L \gg L_{osc}$.

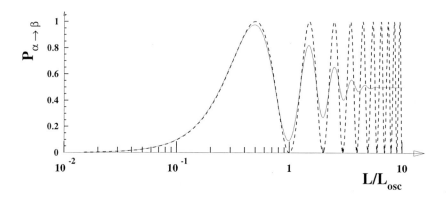

Figure 2.1: Transition probability for $\sin^2 2\theta = 1$ as a function of L/L_{osc}; the
dashed line represents the transition probability (2.8), the solid one the transition
probability averaged over a Gaussian energy neutrino spectrum with mean value
E and $\sigma = E/10$.

The probability expression (2.8) is usually employed in the data analysis
of the neutrino oscillation experiments. The results are always compared
drawing, in the $(\sin^2 2\theta, \delta m^2)$ plane, the confidence domain, i.e. the region
including the oscillation parameters compatible with measurements. If no
indication in favour of neutrino oscillations is found, an upper bound is given
for the transition probability and an exclusion region, lying on the right of
the contour curve, is drawn. Examples of such plots, for some of the most
recent experiments, will be shown in the next Sections.

2.2   Neutrino oscillations in matter

It was first pointed out by Wolfenstein [27] that the passage of neutrinos through dense matter can significantly modify the oscillation pattern in vacuum if ν_e's (or $\bar{\nu}_e$'s) are involved. This effect arises from the coherent neutrino forward scattering which, in addition to the flavour-independent NC interaction amplitude, has a contribution from the CC interactions involving matter electrons. The propagation of neutrinos in matter can be described by adding a potential energy term

$$V(\,\mathrm{eV}) = \sqrt{2}G_F N_e = 1.53 \cdot 10^{-7} \frac{Z}{A}\rho(\,\mathrm{g/\,cm^3}) \qquad (2.9)$$

to the Hamiltonian matrix element $\langle\nu_e|H|\nu_e\rangle$, N_e being the number of electrons per unit volume. In a two-neutrino oscillation scheme, the diagonalization of the Hamiltonian leads to a mixing angle in matter θ_m related to vacuum mixing angle θ by the relation

$$\tan(2\theta_m) = \frac{\delta m^2 \sin(2\theta)}{\delta m^2 \cos 2\theta - 2EV}\,; \qquad (2.10)$$

and to the mass eigenvalues

$$m_m^2 = \frac{1}{2}(m_1^2 + m_2^2 + 2EV) \pm \frac{1}{2}\sqrt{(\delta m^2 \cos 2\theta - 2EV)^2 + (\delta m^2)^2 \sin^2 2\theta} \qquad (2.11)$$

The matter oscillation length is longer than in vacuum and given by

$$L_m = L_{osc}\frac{\delta m^2}{\sqrt{(\delta m^2 \cos 2\theta - 2EV)^2 + (\delta m^2)^2 \sin^2 2\theta}} \qquad (2.12)$$

Eq.(2.10) shows that, even at small vacuum mixing, there could exist neutrino energy or matter density values such that $2EV = \delta m^2 \cos 2\theta$; the denominator vanishes and the mixing angle in matter is $\theta_m = 45°$, corresponding to maximal mixing. This resonant behaviour of the mixing amplitude was first noticed by Mikheyev and Smirnov [28] a few years after the original Wolfenstein formulation and is known as the MSW effect. It is worth stressing that the potential energy term (2.9) changes sign in the case of electron antineutrinos; the denominator never vanishes and no resonance is expected in the oscillation amplitude. As we shall see in the next Section, the resonant MSW effect offers an elegant way to explain the solar neutrino problem even if the mixing angle in vacuum is small.

2.3   Experimental hints

2.3.1   The atmospheric neutrino anomaly

The atmospheric neutrino flux is essentially originated by the leptonic decays of the charged mesons (mostly pions and kaons) produced in the atmosphere by hadronic interactions involving primary cosmic rays. The neutrino production proceeds through the following reaction scheme:

$$\pi^{\pm}, K^{\pm} \to \nu_{\mu}(\overline{\nu}_{\mu}) + \mu^{\pm}$$
$$\downarrow \qquad\qquad\qquad (2.13)$$
$$\mu^{\pm} \to e^{+}(e^{-}) + \nu_{e}(\overline{\nu}_{e}) + \overline{\nu}_{\mu}(\nu_{\mu})$$

The flavour composition of this flux follows immediately from (2.14); denoting neutrino fluxes by (ν), one obtains

$$\frac{(\nu_{\mu}) + (\overline{\nu}_{\mu})}{(\nu_{e}) + (\overline{\nu}_{e})} \simeq 2 \qquad\qquad (2.14)$$

In fact, a lot of factors can affect this relation; for instance, it can be estimated that, at neutrino energies larger than 1 GeV, muons from meson decays start to reach the earth before decaying; this decreases the $\nu_{e}(\overline{\nu}_{e})$ flux and the ratio (2.14) begins to rise [29].

     The largest error in the atmospheric neutrino flux calculations arises from the uncertainty of the order of 15% of the primary cosmic ray spectrum; the modeling of the cosmic ray interaction with light nuclei and of the pion production leads to a $\approx 20\%$ overall uncertainty of the absolute neutrino fluxes. Nevertheless, this major contribution to the systematic error can be cancelled by using the ratio of the muon to electron neutrino flux. This is the reason why the first observable to be measured in recent atmospheric neutrino experiments was the ratio of μ-like events to e-like events (denoted by $(\mu/e)_{data}$. According to their detection technique, these experiments are divided into two classes:

- Čerenkov experiments (such as IMB[3], Kamiokande[2] and, more recently, Super-Kamiokande[30]), where the neutrino target is a large volume of water surveyed by a huge number of photomultipliers to detect the Čerenkov light emitted by neutrino-induced charged particles:

- tracking calorimeter experiments (including Nusex[31], Fréjus[32] and Soudan[4]), measuring both the ionization energy loss and the path of the charged leptons.

Both techniques are sensitive to the direction of the tracks and can thus distinguish between up and down through-going particles. The identification of μ-like events and e-like events is accomplished by separating sharp from diffuse Čerenkov rings (as in the Čerenkov detectors) or through-going tracks from showers (as in the tracking calorimeters). However, the measured $(\mu/e)_{data}$ ratio cannot directly be compared to the corresponding ratio of the atmospheric neutrino fluxes, since both the detector efficiencies and the event selection criteria can affect the expected ratio. As an example, the Monte Carlo simulation of Super-Kamiokande obtains $(\mu/e)_{MC} = 1.5$ in the sample of *fully contained* (FC), sub-GeV events (which are originated by neutrino interactions inside the detector with a fully detector-contained energy deposit and a visible energy $E < 1.3 \, \mathrm{GeV}$); the same ratio rises to 2.83 when considering higher energy events, where either the visible energy is $E > 1.3 \, \mathrm{GeV}$ (multi-GeV events), or the energy deposit is not completely contained in the inner detector (*partially contained* events, PC). The relevant variable to use is then the double ratio

$$R \equiv \frac{(\mu/e)_{data}}{(\mu/e)_{MC}} \tag{2.15}$$

where the expected ratio is obtained by folding the theoretical flux with the neutrino interaction cross sections and the detection efficiencies and by applying the selection criteria.

The first indication of $R < 1$ was reported more than 10 years ago by the IMB collaboration [33]. The next R measurements reported by Čerenkov experiments confirmed this "anomaly" in the atmospheric neutrino flux:

$$R = \begin{array}{lll} 0.61 \pm 0.15 \pm 0.05 & \text{IMB contained} & [3] \\ 0.60^{+0.07}_{-0.06} \pm 0.05 & \text{Kamiokande sub-GeV} & [34] \\ 0.57^{+0.08}_{-0.07} \pm 0.07 & \text{Kamiokande multi-GeV} & [2] \\ 0.63 \pm 0.03 \pm 0.05 & \text{Super-Kamiokande sub-GeV} & [30] \\ 0.65 \pm 0.05 \pm 0.08 & \text{Super-Kamiokande multi-GeV} & [30] \end{array}$$

Measurements with tracking calorimeters gave mixed results. Soudan-2 [4] found $R = 0.61 \pm 0.15 \pm 0.05$, in agreement with the Čerenkov experiments; earlier experiments, such as Fréjus [32] ($R = 1.00 \pm 0.15 \pm 0.08$) and Nusex [31] ($R = 0.96^{+0.32}_{-0.28}$), although affected by larger uncertainties, reported R values compatible with one.

This anomaly is interpreted as strong evidence in favour of neutrino oscillations. Other indications come from the study, performed by Kamiokande and Super-Kamiokande, of the angular dependence of the measured muon and electron neutrino flux; as a matter of fact, with the zenith angle varying from $\vartheta = 0$ (vertically downward going neutrinos) to $\vartheta = \pi$ (vertically upward going neutrinos), the neutrino path length ranges from $\approx 10 \, \mathrm{km}$ to $\approx 13000 \, \mathrm{km}$. Kamiokande [2] observed a zenith angle variation of R for the FC multi-GeV + PC events. The result has been recently confirmed by Super-Kamiokande with much more statistics and with a significant zenith angle variation in the μ-like events, both in the sub-GeV and multi-GeV range. e-like and μ-like events were divided in 5 $\cos \vartheta$ bins and seven momentum bins (from 0.1 to 100 GeV), yielding altogether 70 data points; a statistical analysis was then performed under the assumption of the $\nu_\mu \to \nu_\tau$ oscillation hypothesis, with the overall neutrino flux normalization to be determined by the fit. The best fit gives $\chi^2_{min} = 65.2$ for 67 dof at $\sin^2 2\theta = 1$ and $\delta m^2 = 2.2 \cdot 10^{-3}$ eV^2. The allowed regions in the $(\sin^2 2\theta, \delta m^2)$ at $68, 90, 95\%$ C.L., for $\nu_\mu \to \nu_\tau$ oscillations, are shown in Fig. 2.2a, together with the 90% confidence region previously obtained by Kamiokande; in Fig. 2.2b, the analogous plot obtained by Kamiokande for the $\nu_\mu \to \nu_e$ oscillation is presented.

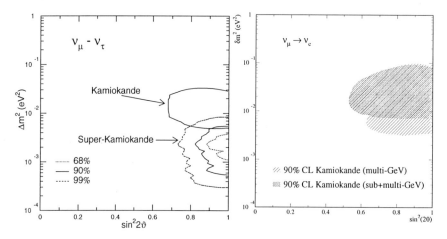

Figure 2.2:   (left) Confidence regions for $\nu_\mu \to \nu_\tau$ oscillations obtained by Kamiokande and Super-Kamiokande; (right) 90% C.L. allowed region by Kamiokande for $\nu_\mu \to \nu_e$ oscillations.

Further evidence in favour of the $\nu_\mu \to \nu_\tau$ oscillation hypothesis was obtained by looking at the baseline L/E dependence of the ratio R and at the up-down asymmetry for the μ-like and e-like events; the results are summarized in Fig. 2.3. The value of the asymmetry for FC and PC multi-GeV μ-like events is $A_\mu \equiv (U-D)/(U+D)_\mu = -0.296 \pm 0.048 \pm 0.01$, which is consistent with the $\nu_\mu \to \nu_\tau$ oscillation hypothesis with $\delta m^2 = 2.2 \cdot 10^{-3}$ eV^2 and $\sin^2 2\theta = 1$. For e-like events, as shown in Fig. 2.2, the asymmetry is a flat function consistent with the no-oscillation hypothesis; the average value for FC and PC multi-GeV events ($A_e = -0.036 \pm 0.067 \pm 0.02$) is compatible with zero.

The study of the angular distribution was extended to higher energy atmospheric neutrinos ($E_\nu \sim 100$ GeV). Since the detection of neutrino interactions inside the detector volume becomes more difficult for higher

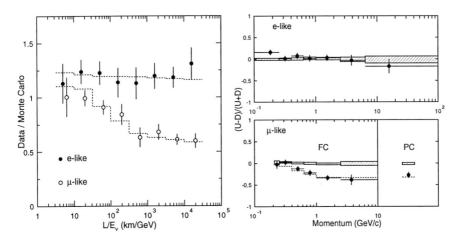

Figure 2.3: (left) Comparison of the (data/MC) ratio vs. neutrino baseline for e-like and μ-like events; the dashed line for μ-like events is the theoretical expectations in the case of $\nu_\mu \to \nu_\tau$ oscillations with $\sin^2 2\theta = 1$ and $\delta m^2 = 2.2 \cdot 10^{-3}\,\mathrm{eV}^2$. (right) Up-down asymmetry vs. momentum for e-like and μ-like events; the hatched region corresponds to the expectations for the case of no oscillations, whereas the dashed line corresponds to the best oscillation hypothesis.

energy neutrinos (because of the steeply falling neutrino spectrum), it is possible to enhance the effective volume of the detector by looking at muons generated in CC interactions of $\nu_\mu(\overline{\nu}_\mu)$'s in the rock below the detector[1]. This technique works only for muons entering the detector from below ("upward going muons") since the neutrino-induced signal from above is dominated by the background of downward cosmic ray muons. Recent measurements of the upward muon flux were reported by the MACRO [35] collaboration. The analysis is based on 479 events, including through-going and stopping muons; the ratio of the number of observed to expected events integrated over all zenith angles is $0.74 \pm 0.036 \pm 0.046$. Fitting the zenith distribution to the $\nu_\mu \to \nu_\tau$ oscillation hypothesis yields a best fit with $\delta m^2 = 2.5 \cdot 10^{-3}\,\mathrm{eV}^2$, $\sin^2 2\theta = 1$ (which is in rather good agreement with the Super-Kamiokande result), although the quality of the fit (maximum χ^2 probability of 5%) is not so good as in Super-Kamiokande.

---

[1]The effective detector volume is then the product of the detector area and the muon range in the rock. For instance, TeV muons have a typical range $\sim 1\,\mathrm{km}$, which leads to a significant increase in the target volume.

2.3.2   The solar neutrino deficit

As all visible stars, the Sun produces energy by the cycle of thermonuclear reactions shown in Fig. 2.4; the net result of hydrogen burning is the transition

$$4p + 2e^- \rightarrow {}^4He + 2\nu_e + Q, \tag{2.16}$$

where $Q = 4m_p + 2m_e - m_{He} = 26.73\,\text{MeV}$ is the total energy release. Hence, energy production is accompanied by the emission of electron neutrinos [36], the neutrino luminosity being a small part ($\approx 2\%$) of the total one. Most of them are low-energy neutrinos coming from the pp reaction; monoenergetic neutrinos with intermediate energy are produced in the electronic capture by ${}^7Be$ and in the pep reaction. Higher energy neutrinos are produced by the ${}^8B$ decay; their flux, even though much smaller than the fluxes of pp, pep and ${}^7Be$ neutrinos, gives the major contribution to the event rate in high threshold experiments. Other small contributions come from the reactions in the CNO cycle; the β^+-decays of nuclei such as ${}^{13}N$, ${}^{15}O$, ${}^{17}F$ are sources of intermediate energy neutrinos (the energy extending up to about 1.7 MeV).

The neutrino spectral shape is known with negligible uncertainties for each source, since it is determined by the weak interaction theory and is practically independent from solar physics. On the other hand, the calculations of the contribution of each reaction to the integral neutrino flux must rely on a solar evolution model and the resulting uncertainties represent one of the main problems in explaining the experimental results. The Standard Solar Model (SSM), developed about 20 years ago by Bahcall [36], is based on several assumptions (about the thermal and hydrostatic equilibrium and the energy transport inside the Sun) and requires a detailed knowledge of many parameters (such as the total luminosity, the core and the surface temperature, the hydrogen and helium content in the core). Moreover, little information is directly available in the low energy range for the cross section of the nuclear reactions involved in the neutrino emission. The SSM underwent several revisions in the last years, motivated by the disagreement between observations and predictions.

The first indication of a deficit of the neutrino counting rate to the predicted one was pointed out about 30 years ago by the pioneering ${}^{37}Cl$-based radiochemical experiment carried out by Davis *et al.* in the Homestake mine [37]. The experimental technique is based on an extraction of the radioactive ${}^{37}Ar$ produced by solar neutrinos through the reaction

$$\nu_e + {}^{37}Cl \rightarrow e^- + {}^{37}Ar, \tag{2.17}$$

the one originally proposed by Pontecorvo. Since the energy threshold of reaction (2.17) (0.81 MeV) is above the end-point of the pp neutrino spectrum,

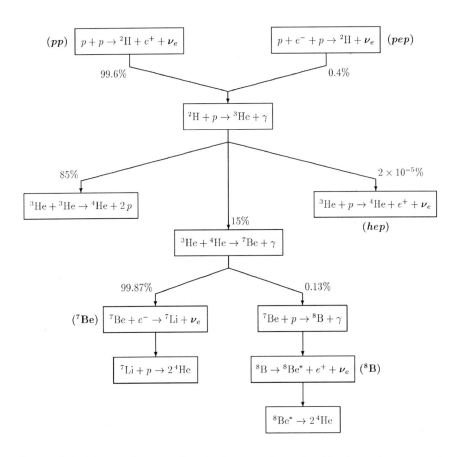

Figure 2.4: Main nuclear reactions in the pp cycle involved in thermal energy and neutrino production inside the sun; their contribution to the overall neutrino flux is also listed.

the main contribution to the counting rate comes from ^{8}B and ^{7}Be neutrinos. The result, expressed in SNU units (1 SNU $\equiv 10^{-36}$ events atom^{-1} s^{-1}), is shown in Tab. 2.2; it turns out to be 1/3 of the Bahcall-Pinsonneault predictions based on the standard solar model (SSM) [38].

Other subsequent solar neutrino experiments reported evidence of a solar neutrino deficit, as seen in Tab. 2.2. The radiochemical experiments GALLEX [40] and SAGE [41] used a Gallium target to detect electron neutrinos via the reaction

$$\nu_e + {}^{71}\text{Ga} \rightarrow e^- + {}^{71}\text{Ge} \tag{2.18}$$

Since the threshold of the process (2.18) is 0.233 MeV, these experiments had the possibility to detect pp neutrinos, thus providing the first experimental

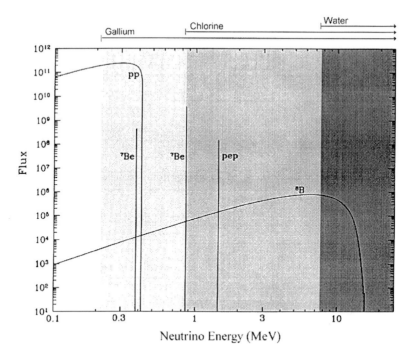

Figure 2.5: Neutrino energy spectrum for the various *pp* cycle reactions; the sensitivity region for the different types of experiments is also indicated.

Table 2.2: Summary of the results obtained by the solar neutrino experiments; each measurement is compared with the predictions based on the standard solar model [38]. The flux is expressed in SNU units for the radiochemical experiments and in terms of the ^{8}B flux (in units of $10^{-6}\,\mathrm{cm}^{-2}\mathrm{s}^{-1}$) for the water Čerenkov experiments.

Experiment	Measured	Expected	meas./exp.
Homestake[39]	$2.56 \pm 0.16 \pm 0.16$	$7.7^{+1.2}_{-1.0}$	$0.33^{+0.06}_{-0.05}$
GALLEX[40]	$77.5 \pm 6.2^{+4.3}_{-4.7}$	129^{+8}_{-6}	0.60 ± 0.07
SAGE[41]	$66.6^{+6.8+3.8}_{-7.1-4.0}$	129^{+8}_{-6}	0.52 ± 0.07
Kamiokande[42]	$2.80 \pm 0.19 \pm 0.33$	$5.15^{+1.0}_{-0.7}$	0.54 ± 0.07
Super-Kamiokande[43]	$2.44 \pm 0.05^{+0.09}_{-0.07}$	$5.15^{+1.0}_{-0.7}$	$0.47^{+0.09}_{-0.07}$

proof of the thermonuclear origin of the solar energy production. However, the measured event rate is about one half of the SSM predictions, the difference being of the order of seven standard deviations. In order to gain confidence in these rates, both detectors were calibrated by using an intense

[51]Cr neutrino source [40, 41]. The results rule out the presence of unexpected systematic errors associated with the extraction technique at the 10% level in both experiments.

In Kamiokande [42] and Super-Kamiokande [30], solar neutrinos are detected through the observation of the Čerenkov light emitted by the recoil electron in the elastic scattering

$$\nu_e + e^- \rightarrow \nu_e + e^- \tag{2.19}$$

The direction of the sun is measured by using the forward-peaked angular distribution of the recoil electron. The energy threshold of the water Čerenkov detectors is higher than in other solar neutrino experiments ($E_{thr} \sim 7\,\mathrm{MeV}$ in Kamiokande, $E_{thr} \sim 6\,\mathrm{MeV}$ in Super-Kamiokande); therefore only ^{8}B can be detected. Also in this case (see Tab. 2.2) the neutrino event rate is about one half of the SSM predictions.

The discrepancy between the experimental results and the predictions is what is called the "solar neutrino problem". Several attempts to slightly modify the input parameters of the SSM were made in order to reduce this discrepancy [44]. Recently the SSM has been successfully tested by comparing its predicted value for the sound speed in the inner layers on the sun with precise helioseismological measurements [38, 45]. Model-independent information on the neutrino fluxes can be extracted from the experimental data by using minimal constraints on the total solar luminosity [46]; for instance, it is possible to obtain a lower bound on the Gallium event rate of (77 ± 2) SNU, which is just compatible with the combined result of the Gallium experiments (see Tab. 2.2). Such an experimental result could be explained if only pp neutrinos were emitted by the Sun, which is incompatible with the helioseismological data. Similarly, by combining all the experimental results, one obtains a strong suppression of the fluxes of intermediate energy (pep, ^{7}Be and CNO) neutrinos with respect to SSM predictions, which is in disagreement with any solar model constrained by the helioseismological data.

So, if we believe in the experimental results, the only possible explanation left is the one of neutrino oscillations [47], either through vacuum or through MSW resonant transition in matter. For neutrinos propagating through the Sun, the density varies along the trajectory from a value higher than $100\,\mathrm{g/cm^3}$ in the core to much less than $1\,\mathrm{g/cm^3}$ at the surface; also the Z/A ratio varies as a function of the hydrogen abundance. Hence the Hamiltonian of the neutrino state is a function of time; the ideal case of constant density discussed in the previous Section is a good approximation ("adiabatic" approximation) if the density variation over an oscillation length is negligble, i.e.

$$\frac{1}{\rho}\frac{d\rho}{dr}\lambda_m \ll 1, \tag{2.20}$$

r being the distance from the Sun centre. The neutrino state is described as the superimposition of mass eigenstates with slowly varying eigenvalues and a mixing angle. If the neutrino energy is high enough that the condition $2EV > \delta m^2 \cos 2\theta$ holds, then $\theta_m > 45°$ and the dominant mass eigenstate is ν_2 (i.e., the one projected along ν_μ for small vacuum mixing). If the adiabatic approximation still holds at resonance, where λ_m is maximal, then the transition probability $\nu_2 \rightarrow \nu_1$ is negligible and the dominant mass eigenstate is still ν_2 as the neutrino emerges from the Sun. Thus the initial ν_e state can be rotated to ν_μ, even for small mixing angles, as a result of the MSW effect. We recall that this condition is satisfied only by neutrinos above a critical energy which depends on the mixing parameters. Since the detection reactions have different energy thresholds, an analysis of the solar neutrino data in terms of matter-enhanced oscillations provides different sensitivity domains to the oscillation parameters for the various experiments.

The allowed regions in the $(\sin^2 2\theta, \delta m^2)$ plane, obtained by the fit of the measured event rates listed in Tab. 2.2 by using the Bahcall-Pinsonneault SSM [38]), are shown in Fig. 2.6. In the case of vacuum $\nu_e \rightarrow \nu_{\mu,\tau}$ oscillations,

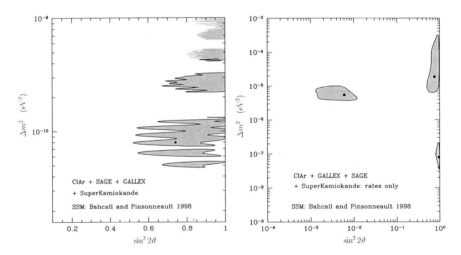

Figure 2.6: Regions allowed (shadowed areas) to $\nu_e \rightarrow \nu_{\mu,\tau}$ oscillations at 99% C.L. for vacuum (left) and MSW (right) transitions.

this region extends over a large interval around the best-fit values $\sin^2 2\theta = 0.75$ and $\delta m^2 = 8 \cdot 10^{-11}$. The 99% confidence domain for MSW $\nu_e \rightarrow \nu_{\mu,\tau}$

oscillations presents instead, for the reasons explained above, three disjoint regions:

- a small mixing angle region with best-fit point at $\sin^2 2\theta = 6 \cdot 10^{-3}$ and $\delta m^2 = 5 \cdot 10^{-6} \, \text{eV}^2$:

- a large mixing angle, large mass region with best-fit point at $\sin^2 2\theta = 0.76$ and $\delta m^2 = 2 \cdot 10^{-5} \, \text{eV}^2$:

- a large mixing angle, low mass region with best-fit point at $\sin^2 2\theta = 0.96$ and $\delta m^2 = 8 \cdot 10^{-8} \, \text{eV}^2$.

In conclusion, the results of all the solar neutrino experiments provide strong evidence in favour of neutrino oscillations. Up to now only information about the observed rates were used in fitting the experimental data. More detailed model-independent information on the neutrino oscillation probability will be obtained by looking for a possible distortion of the electron neutrino spectrum. Apart from Super-Kamiokande, a measurement of this spectrum will be perhaps provided by the SNO experiment [48], a Čerenkov detector consisting of a 1-kton heavy water neutrino target. In addition to the elastic scattering, solar neutrinos will be real-time observed through the CC and NC interactions on Deuterium:

$$
\begin{aligned}
\nu_e + \text{d} &\rightarrow e^- + \text{p} + \text{p} &&(CC) \\
\nu_l + \text{d} &\rightarrow \nu_l + \text{p} + \text{n} \quad (l = e, \mu, \tau) &&(NC)
\end{aligned}
\qquad (2.21)
$$

Since the energy threshold for the recoil electron is about $5 \, \text{MeV}$ and the neutrino energy threshold for the NC reaction is $2.2 \, \text{MeV}$, only ^8B can be observed. The energy of recoil electrons from CC reactions (whose cross section is about 10 times as large as the electron scattering) can be reconstructed as in Kamiokande and Super-Kamiokande. A distortion of this spectrum with respect to the calculated one (in the absence of neutrino oscillations) will provide model-independent evidence of neutrino oscillations. The measurement of NC interactions, which are independent of the neutrino flavour, will allow SNO to determine the total neutrino flux from ^8B; estimations based on the SSM give an NC interaction rate $\sim 3 \cdot 10^3$ events/y. An agreement of the NC event rate with expectations and evidence of a deficit in the NC/CC event rate will definitely prove that neutrino oscillations explain the solar neutrino problem.

An interesting tool to search for possible effects due to matter oscillations in the earth is the "day-night asymmetry". The passage of neutrinos through the earth can cause a regeneration of ν_e's and a consequent increase of the neutrino counting rate; this effect is larger during the night, when the neutrino path in the earth to the detector is longer than during the day. The size of such an asymmetry is significant only for large values of the mixing angle. The preliminary value given by Super-Kamiokande ($-0.023 \pm 0.020 \pm 0.014$) is still compatible with zero, thus suppressing at present the compatibility of the data with the large mixing angle solution [43].

The Borexino experiment [49] is designed to detect low-energy neutrinos ($E_{thr} = 0.25 \, \mathrm{MeV}$) in real-time through elastic scattering in a 300 ton, liquid scintillator unsegmented target. The event rate predicted by SSM for a fiducial volume of 100 tons is $\sim$ 50 events/d, mostly induced by the ^{7}Be neutrinos. It will then be possible to measure this flux and check whether it is suppressed with respect to the one predicted by the SSM, as suggested by the results of current experiments.

2.3.3 The LSND result

The LSND experiment [50] is a short baseline neutrino oscillation experiment at the Los Alamos Meson Physics Facility (LAMPF). Neutrinos are produced by a 800 MeV proton beam hitting a water target so as to produce pions (the proton energy is below the threshold for producing kaons). Most of the π^+'s are stopped in the beam target and decay into muons, which stop and decay in the target as well. The $\bar{\nu}_\mu$'s from the decay at rest of positively charged muons allow to investigate $\bar{\nu}_\mu \to \bar{\nu}_e$ oscillations. The π^+ selection is to suppress the contamination of the beam from $\bar{\nu}_e$'s, thus making possible an appearance test of this oscillation channel; little contamination ($< 10^{-3}$) arises from the π^- decay chain. The energy spectrum of the $\bar{\nu}_\mu$ flux (which extends up to 52.8 MeV) is very well known, being determined by the kinematics of the muon decay. The electron antineutrinos are detected by the inverse β-decay reaction

$$\bar{\nu}_e + \mathrm{p} \to \mathrm{e}^+ + \mathrm{n}, \tag{2.22}$$

which, as discussed in the next Chapter, is by far the most efficient in this neutrino energy range. The $\bar{\nu}_e$ signature is a coincidence between the prompt e^+-signal and the delayed $\gamma(2.2 \, \mathrm{MeV})$ from the capture reaction

$$\mathrm{n} + \mathrm{p} \to \mathrm{d} + \gamma \tag{2.23}$$

The neutrino target consists of a cylindrical (8.3 m long, 5.7 m wide) tank filled with a mineral-oil-based scintillator. An array of photomultipliers detect both the scintillation and the Čerenkov light; this is used for identifying the positron signals (for which the Čerenkov light is $\sim$ 1/5 of the scintillation light) from the signals associated with the recoil protons (whose velocity is below the Čerenkov threshold) by neutron scattering, which represent the most dangerous source of background for $\bar{\nu}_e$ detectors.

LSND reported an excess of e^+ events ($51.0^{+20.2}_{-19.5} \pm 8.0$) with energies from 20 MeV to 60 MeV, corresponding to a $\bar{\nu}_\mu \to \bar{\nu}_e$ transition probability of $(3.1 \pm 1.2 \pm 0.5) \cdot 10^{-3}$ [50]. The allowed region in the $(\sin^2 2\theta, \delta m^2)$ plane is shown in Fig. 2.7a. More recently, the small fraction ($\sim 3.4\%$) of ν_μ's produced by π^+ decays in flight was used to test the $\nu_\mu \to \nu_e$ oscillation channel. The ν_e's are detected through the chain reaction

$$\nu_e + {}^{12}C \to {}^{12}N_{\text{g.s.}} + e^-$$
$$\downarrow \qquad\qquad\qquad (2.24)$$
$$ {}^{12}N_{\text{g.s.}} \to {}^{12}C + e^+ + \nu_e $$

the lifetime of the ${}^{12}N_{\text{g.s.}}$ being 15 ms. 5 events were observed against an expected background of 0.5 events [51], thus yielding new evidence in favour of neutrino oscillations; the allowed region is shown in Fig. 2.7b.

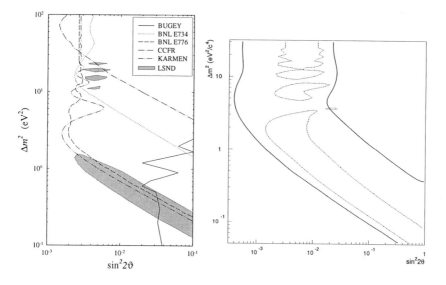

Figure 2.7: (left) Allowed regions (shadowed area) at 90% C.L. by LSND to $\bar{\nu}_\mu \to \bar{\nu}_e$ oscillations superimposed on the exclusion contours of Bugey3, BNL E734, BNL E776, CCFR and KARMEN. (right) Allowed regions for $\nu_\mu \to \nu_e$ oscillations (π^+ decay in flight analysis) by LSND at 95% (solid line) compared with the 99% C.L. region allowed by the decay at rest analysis (dotted line).

The region of oscillation parameters indicated by the LSND results is currently explored by the KARMEN experiment [52], located at the ISIS pulsed spallation neutron facility at the Rutherford Laboratories. The positive pions, produced by protons hitting the beam target, decay at rest producing an equal number of ν_μ's, ν_e's and $\bar{\nu}_\mu$'s; $\bar{\nu}_e$'s are then searched for at an 18 m average distance from the target. The $\bar{\nu}_e$ detection is also based on the inverse β-decay reaction and the associated signature is provided, as in the LSND case, by the coincidence of the e^+ and the neutron-capture signal. The time structure of the neutrino beam is used to identify the neutrino-induced reactions from the cosmic-ray associated background. No event has been observed so far, with an expected background of 2.88 ± 0.13 events; the resulting exclusion plot is superimposed on the LSND plot in Fig. 2.7. At present the null result by KARMEN excludes part of the LSND-allowed region; in one or two years KARMEN sensitivity will completely cover that region, thus confirming or excluding the LSND result.

It is quite difficult to explain the LSND result, as well as the solar neutrino deficit and the atmospheric neutrino anomaly, in terms of oscillations involving three neutrino flavours; the δm^2 values required by the data are very different from each other whereas only two independent mass differences can be defined with three neutrinos. At least one sterile neutrino is thus required to accomodate all the experimental data, which would provide new evidence for physics beyond the Standard Model [53]. This is the reason why a number of experiments were recently proposed at accelerators to test the LSND result with improved sensitivity; we recall BooNE [54] at Fermilab (which is already approved) I-216 [55] at CERN and NESS [56] at the European Spallation Source.

2.4 Experiments at accelerators

2.4.1 The present: CHORUS and NOMAD

We have just mentioned the two major searches (LSND and KARMEN) for $\nu_\mu \rightarrow \nu_e$ oscillations. Two experiments are presently operating at the CERN-SPS wide-band neutrino beam to search for $\nu_\mu \rightarrow \nu_\tau$ oscillations: CHORUS [57] and NOMAD [58]. They explore the high mass difference $(\delta m^2 \gtrsim 1\,\text{eV}^2)$ region, where a signature is expected if neutrino masses significantly contibute to the hot component of the dark matter.

Both experiments are designed to detect τ's from ν_τ interactions in the detector target; the neutrino beam energy ranges from 10 to 40 GeV, above the energy threshold for τ production. The sensitivity goal is a ν_τ / ν_μ ratio of $\sim 10^{-4}$, about three order of magnitudes larger than the estimated ν_τ contamination in the neutrino beam from the decay $(D_s^\pm \rightarrow \tau^\pm \nu_\tau (\bar{\nu}_\tau))$ of the D_s mesons produced by the primary proton beam. The observation of τ production could therefore result only from $\nu_\mu \rightarrow \nu_\tau$ oscillations.

The CHORUS detector is located at a distance $\sim 820\,\text{m}$ from the proton target. The τ detection technique is based on the identification of the short-lived τ-decay in nuclear emulsion. The excellent space resolution ($< 1\mu$m) achieved by the emulsion technique serves to localize the kink due to the τ-decay, whose distance from the interaction vertex is of the order of 1 mm. The explored decay branches are:

$$\begin{aligned}
\tau^{\pm} &\rightarrow \mu^{-}\overline{\nu}_{\mu}\nu_{\tau} & (BR = 17.6\%) \\
&\rightarrow h^{-}(n\pi^{0})\nu_{\tau} & (BR = 49.8\%) \\
&\rightarrow 2\pi^{-}\pi^{+}(n\pi^{0}) & (BR = 14.9\%)
\end{aligned} \qquad (2.25)$$

The detector core consists of an emulsion target (total mass ~ 800 kg), followed by a scintillation fibre tracking array coupled to a high-resolution calorimeter and a muon spectrometer for the identification of the τ-decay products. If the event topology satisfies one of the decay branches in (2.25), the tracks are followed back to the exit point from the emulsion target for further scanning. Simple kinematical criteria are applied to reject the large number of events due to ν_{μ} CC or NC interactions satisfying the first selection. For the selected sample (reduced by more than one order of magnitude with respect to the total) a computer-assisted microscope is used to scan the emulsion sheets back to the primary interaction vertex, according to the procedure illustrated in Fig. 2.8.

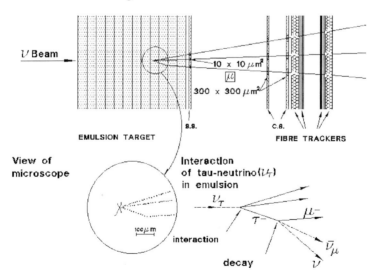

Figure 2.8: Expected configuration of a typical ν_{τ}N $\rightarrow \tau^{-}$X CHORUS event in the emulsion and in the scintillation fibre tracker.

The CHORUS experiment started taking data in May 1994 and stopped in 1997; the total exposure amounts to $5.06 \cdot 10^{19}$ protons on target and $\sim 2 \cdot 10^{6}$ emulsion triggers (half of which still have to be scanned). Since no ν_{τ}

candidate event was found in the scanned sample, an upper limit at 90% C.L.
$P_{\mu\tau} < 4.0 \cdot 10^{-4}$ was set to the oscillation probability; for $\delta m^2 > 40\,\text{eV}^2$ this
corresponds to the limit $\sin^2 2\theta < 8 \cdot 10^{-4}$ [57].

In the NOMAD experiment, located just downstream from the CHORUS
detector, the identification of the τ^- production and decay is uniquely based
on kinematical selection criteria associated with the missing energy from
outgoing neutrinos. It is then possible to study further decay branches, such
as $\tau^- \rightarrow e^- \bar{\nu}_e \nu_\tau$; the main background for this process, resulting from the
ν_e CC interactions in the target volume, is rejected by means of kinematical
constraints. The experimental set-up consists of a drift chamber system (also
acting as the neutrino target, mass ~ 2.5 tons) immersed in a 0.4 T horizontal
magnetic field for a high-resolution measurement of charged hadrons and
muons. The drift chamber system is coupled to:

- a Transition Radiation Detector for electron identification;

- a lead-glass electromagnetic calorimeter with an excellent energy resolution
  $(\sigma/E = 0.01 \oplus 0.04/\sqrt{E(\text{GeV})}$ for electrons and photons);

- a sampling hadronic calorimeter made of a multi-layer iron-scintillator
  sandwich;

- a large-area muon chamber system for muon identification.

Since the number of ν_τ events is compatible with the expected background,
an upper limit $P_{\mu\tau} = 2.1 \cdot 10^{-3}$ was set to the oscillation probability; this
corresponds to an asymptotic limit $\sin^2 2\theta < 4.2 \cdot 10^{-3}$.

As a consequence of these preliminary negative results, the possibility
that massive neutrinos significantly contribute to dark matter is ruled out,
unless the neutrino mixing is very small.

2.4.2   Long baseline projects at accelerators

The explanation of the atmospheric neutrino anomaly in terms of neutrino
oscillations can be checked with long-baseline neutrino oscillation experiments. The main advantage of these experiments over atmospheric neutrino
experiments is that the neutrino beam is better characterized, both in energy
(as in the case of reactors) and in time structure (as in the case of accelerators). Moreover, the neutrino fluxes also have higher intensity, so that many
possible systematic effects are under better control. In the case of higher
energy Fermilab and CERN beams, it may also be possible to directly search
and detect ν_τ's, if these neutrinos are produced in the oscillations. Let us
deal now with the proposed searches at accelerators, deferring an analysis of
reactor experiments to the following Sections.

The first operating search is the K2K experiment [59], with a distance of 235 km from KEK to Super-Kamiokande. The average neutrino energy (1.4 GeV) of the beam makes the experiment sensitive to ν_μ disappearance and $\nu_\mu \to \nu_e$ transitions with $\delta m^2 \gtrsim 2 \cdot 10^{-3}$. A near 1 kton water detector will be placed at a distance of 1 km from the beam dump in order to monitor the initial flux and energy spectrum of ν_μ's. It has been taking data since last January; the sensitivity achievable by one year exposure is plotted in Fig. 2.9, superimposed on the region allowed by Super-Kamiokande for $\nu_\mu \to \nu_\tau$ oscillations.

The MINOS experiment [60] is under construction in the Soudan underground laboratory, at a distance of 730 km from the neutrino factory at Fermilab. The neutrino beam will be produced by protons from the new Main Injector and will have an average energy ~ 10 GeV. The far detector consists of an 8 kton sampling calorimeter with alternating magnetized iron and scintillator slabs; it will work in conjunction with a near detector of similar design located at a distance of 1 km from the proton target. With such a mass and wide-band beam, a number of $20000 \nu_\mu$ CC events per year is expected. Moreover, the ν_μ beam energy serves to test the ν_μ disappearance as well as the $\nu_\mu \to \nu_{e,\tau}$ appearance, with the possibility of distinguishing the different channels. The goal of the experiment is to gain sensitivity to $\nu_\mu \to \nu_x$ oscillations for $\delta m^2 \gtrsim 10^{-3}$ eV^2 at maximum mixing, so as to cover a big part of the Super-Kamiokande allowed region, as shown in Fig. 2.9. Data taking is expected to start in 2002.

A variety of proposals (such as OPERA [61], NOE [62], AQUA-RICH [63], NICE [64]) exists for long-baseline oscillation searches at the Gran Sasso underground laboratory by using the planned neutrino beam facility to be built at CERN (730 km away). The objective is to study the $\nu_\mu \to \nu_x$ oscillations, in the $\delta m^2 \gtrsim 10^{-3}$ range indicated by the solution to the atmospheric neutrino anomaly, The ICARUS experiment [65] is scheduled to start in the year 2000 with a first TPC module filled by 600 ton liquid argon; a future extension of the detector to 2.4 Kton total mass is also planned. This detector, which is also sensitive to atmospheric and solar neutrinos, will allow to detect long-baseline $\nu_\mu \to \nu_\tau$ oscillations; since the average energy of the proposed beam is rather high (about 25 GeV) in order to allow the ν_τ detection through CC interactions, the goal of the experiment is to be sensitive to $\delta m^2 \gtrsim 10^{-3}$.

2.5   Experiments at nuclear reactors

Nuclear reactors were the first source to be used to search for neutrino oscillations and, in general, for systematic studies of neutrino properties (it is enough to think of the experiment by Reines & Cowan). A summary of

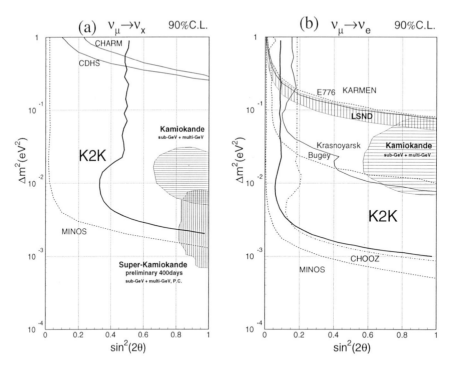

Figure 2.9: Expected sensitivity to neutrino oscillations for future LBL searches at accelerators, compared with the Kamiokande and Super-Kamiokande allowed regions for atmospheric neutrinos and the limits from reactor and SBL accelerator experiments. Fig.(a): $\nu_\mu \to \nu_x$ test. Fig.(b): $\nu_\mu \to \nu_e$ test.

the main neutrino experiments at reactors is given in Tab. 2.3. A more detailed description of the reactor as a neutrino source will be given in the next Chapter. It is now worth recalling that, because of the low neutrino energy ($\langle E_\nu \rangle \sim 3\,\mathrm{MeV}$), only disappearance oscillation tests are possible at reactors.

2.5.1   Previous experiments

In past experiments (i.e. up to the eighties), the knowledge of the reactor neutrino source (accurate at $\approx 10\%$ at that time) was the factor that limited the sensitivity of the neutrino oscillation tests to small mixing angles. This is the reason why, as shown in Tab. 2.3, oscillation tests were performed by comparing the neutrino event rate at different distances (up to $\sim 100\,\mathrm{m}$) from the reactor. This approach had the advantage of ruling out the systematic uncertainty associated with the absolute neutrino flux, whose uncertainty was

Table 2.3: Parameters and results of the experiments carried out at reactors. Among these, Palo Verde is the only one currently taking data; Kamland is expected to start operating in January 2001. The δm^2 sensitivity limit, wherever omitted, is obtained at 90% CL.

experiment site	target mass(kg)	full power (GWth)	reactor distance(m)	δm^2(eV^2) ($\sin^2 2\theta = 1$)
Savannah River[66]	260	2.2	18, 24	0.1
Gösgen[6]	320	2.8	38, 46, 65	0.019
Rovno[67]	200	1.4	18, 25	0.014 (68% CL)
Krasnoyarsk[68]	600	1.6	58, 231	$7.5 \cdot 10^{-3}$
Bugey[7]	1200	2.8	15, 40, 95	0.01
Palo Verde[69]	11300	10.9	740, 850	$2 \cdot 10^{-3}$
Chooz[70]	5500	8.5	1000, 1100	$9 \cdot 10^{-4}$
Kamland[71]	10^6	127	$\approx 1.5 \cdot 10^5$	10^{-5} (goal)

much larger than the statistical error (only a few percent). Let us review the most recent short-baseline experiment at reactors.

Gösgen

A search for neutrino oscillations was performed at the 2800 MW$_{th}$ nuclear power station in Gösgen (Switzerland) [6], providing an isotropic flux of $5 \times 10^{20} \bar{\nu}_e \, \mathrm{s}^{-1}$. The detection technique was based on the inverse β-decay reaction; the detector (approximately 1 m^3 total size) consisted of 30 cells arranged in 5 planes and filled with a proton-rich liquid scintillator (NE235C), alternated by 4 multi-wire proportional chambers filled with ^{3}He. The cells worked as a target for the $\bar{\nu}_e$, as a calorimeter for the generated positrons and at the same time as a moderator for the emerging neutrons; these are detected in the wire chambers through the reaction

$$\mathrm{n} +^3 \mathrm{He} \rightarrow^3 \mathrm{H} + \mathrm{p} + 765 \, \mathrm{KeV} \tag{2.26}$$

delayed by $\approx 150 \, \mu\mathrm{s}$ (including the moderation and the diffusion time) with respect to the positron. The signature for $\bar{\nu}_e$ events required a proper spatial correlation as well as a time coincidence within 250 μs between the positron and the neutron signal.

The $\bar{\nu}_e$ energy spectrum was measured at three different distances (37.9, 45.9 and 64.7 m) from the reactor core; about 10^4 events were recorded at each detector position. The analysis of the detected spectra showed the consistency of the data with the absence of neutrino oscillations; the resulting bounds at 90% C.L. on the oscillation parameters are $\delta m^2 < 0.019\,\mathrm{eV}^2$ at full mixing and $\sin^2 2\theta < 0.21$ for $\delta m^2 > 5\,\mathrm{eV}^2$.

Bugey

More stringent bounds on the oscillation parameters were obtained by the experiment carried out at the nuclear power station in Bugey (France) [7], where 5 pressurized water reactors, similar to the one at Gösgen, are operating. A measurement of the $\bar{\nu}_e$ spectra was performed at three different source-detector distances $(15, 40,$ e $95\,\mathrm{m})$ by using three identical modules filled by ^{6}Li-doped liquid scintillator. The signature for $\bar{\nu}_e$ events consisted of a double coincidence of a prompt signal due to the positron, and a delayed (by $30\,\mu s$ on average) signal associated with the neutron capture reaction

$$\mathrm{n} + {}^6\mathrm{Li} \to {}^4\mathrm{He} + {}^3\mathrm{H} + 4.8\,\mathrm{MeV} \tag{2.27}$$

Because of the strong scintillation quenching, the visible energy corresponding to the capture energy reduces to $530\,\mathrm{KeV}$; an additional tool for neutron identification was obtained by the good pulse shape discrimination property of the scintillator.

The $\bar{\nu}_e$ spectra recorded at each detector position were in agreement with the spectra computed for the no-oscillation case; the exclusion plot at 90% C.L. is shown in Fig. 2.9 superimposed on the long baseline search contour. The reactor-detector distance, longer than in Gösgen, allowed Bugey to set a more stringent bound $\delta m^2 < 0.01\,\mathrm{eV}^2$ to the mass difference values; on the other hand, the larger collected statistics $(1.2 \cdot 10^5$ at the nearest site, $3 \cdot 10^4$ at 40 m) implied an upper bound $\sin^2 2\theta < 0.14$ at large δm^2 values.

2.5.2   Towards long baselines: Chooz and Palo Verde

Past experiments at reactors worked at distances of the order of tens or hundreds of metres, so exploring a δm^2 region down to values slightly below $10^{-2}\,\mathrm{eV}^2$ at full mixing (see Tab. 2.3). The Chooz experiment [5] is the first long baseline search $(L/E \gtrsim 300\,\mathrm{m/\,MeV})$ utilizing artificial neutrino sources ever operated. The distance of the detector from the reactor is about one order of magnitude larger than previous experiments, which improves the sensitivity to a region of mass difference values $(\delta m^2 \gtrsim 10^{-3}\,\mathrm{eV}^2)$ never explored before, where hints at neutrino oscillations come from the atmospheric neutrino results. The first result obtained by Chooz [70] already excludes atmospheric $\nu_\mu \to \nu_e$ oscillations in the sensitivity mass domain with $\sin^2 2\theta \gtrsim 0.2$. The last results from Chooz, which is the object of this thesis, will put more severe constraints on the mixing of electron neutrinos with other flavours. We refer to the following Chapters for a detailed description of the experiment and data analysis.

The Palo Verde experiment [69] has been taking data since May 1998 at the homonymous power station in Arizona (US). The experimental site is located at an average distance of 800 m from three reactors providing $10.9\,\mathrm{GW}_{th}$ total power. The experiment uses Gd-doped liquid scintillator like the Chooz experiment and the same delayed coincidence technique for neutron identification. The detector is installed in an underground laboratory at a depth of 32 MWE, which is sufficiently deep to eliminate the hadronic component of cosmic radiation, but too shallow to suppress the high energy neutron flux induced by cosmic muon spallations. The detector design aims at reducing this major source of background; it consists therefore of a segmented liquid scintillation calorimeter, made of 66 acrylic cells of $9 \times 0.25 \times 0.25\,\mathrm{m}^3$ for a total fiducial volume of about 12 tons, surrounded by a 1 m water buffer and an active liquid scintillator veto. The positrons from $\overline{\nu}_e$ interactions are identified by requiring a prompt triple coincidence of a positron signal, due to ionization loss in a cell, with the associated 511 KeV annihilation quanta in neighbouring back to back cells. The positron signal is furthermore required to have proper spatial and temporal correlation with the delayed neutron capture signal.

The Palo Verde collaboration obtained a prelimanary result based on an early data set of 39 days with full power and 33 days with only 2 reactors on, corresponding to 71% of the full-power flux. The event rates for the two periods were found to be $(39.1 \pm 1.0)\,\mathrm{d}^{-1}$ and $(32.6 \pm 1.0)\,\mathrm{d}^{-1}$ respectively, with a difference of $(6.4 \pm 1.4)\,\mathrm{d}^{-1}$ attributed to the shut down reactor and a signal-to-background ratio of 1.2; by applying the detection efficiencies, this corresponds to $77 \pm 17 \pm 11$ daily $\overline{\nu}_e$ interactions, in agreement with the rate of $(59 \pm 2)\,\mathrm{d}^{-1}$ predicted for the no-oscillation case. The resulting exclusion plot, together with the first Chooz exclusion plot and with the parameter range allowed by the Kamiokande atmospheric results, is shown in Fig. 2.10.

2.5.3   The future: KamLAND

The KamLAND experiment [71] will use the Kamiokande site to perform an ultra-long baseline oscillation search with sensitivity to mass values improved to $\delta m^2 \gtrsim 10^{-5}\,\mathrm{eV}^2$, so as to test the large mixing angle MSW solution of the solar neutrino problem. It will measure the neutrino flux from several reactors at distances exceeding 100 km; the main constribution (39.1%) to this flux will come from the Kashiwazaki reactor, which is 160 km from the site and provides $24.4\,\mathrm{GW}_{th}$ total power. A schematic view of the detector,

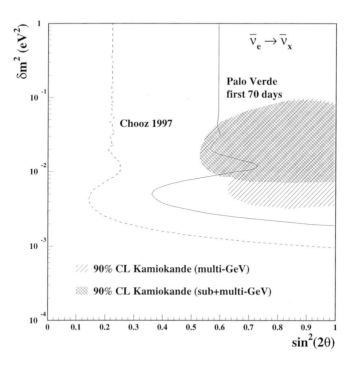

Figure 2.10: Exclusion plot at 90% C.L. for the 72-day run of the Palo Verde experiment; the Chooz 1997 results and the Kamiokande allowed region are also shown.

whose building has been going on since 1998, is shown in Fig. 2.11. The neutrino target will consist of a spherical transparent balloon filled with 1000 tons of non-doped liquid scintillator; the target will be surrounded by a concentrical, 2.5 m thick, mineral oil shielding buffer, in order to maximize the light collection by the photomultipliers. Neutrinos will be detected by the inverse β-decay reaction accompanied by the delayed neutron capture on proton (2.23). The high capture time (175 μs) and low energy of the emitted γ (2.2 MeV) demand good scintillator timing resolution (for position reconstruction) and good pulse shape discrimination properties (for neutron identification) in order to keep the background at acceptable levels. The expected neutrino signal is $\sim$ 2 events per day in case of no oscillations; the background (mainly accidental) is estimated to be ~ 0.1 events per day. One of the main background sources is the β-activity associated with the spontaneous fissions of Uranium and Thorium nuclei present in the earth.

This background is estimated to contribute with $50 \div 150$ events per year, thus implying a signal to noise ratio $\sim 10/1$. Such a background can be reduced by applying a selection on the positron energy, which does not exceed 2.4 MeV in the case of "geophyisical" neutrinos.

Data taking is scheduled to start in 2001.

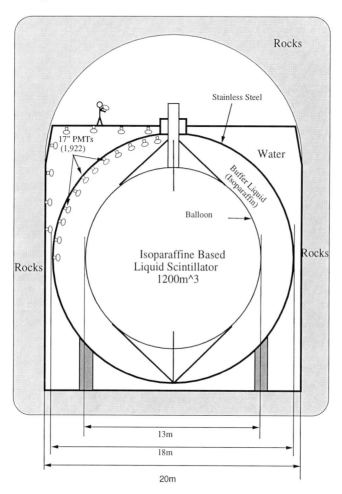

Figure 2.11: Schematic view of the KamLAND detector.

Part II

The Chooz experiment

Chapter 3

The neutrino source

The Chooz experiment, just like the other searches for neutrino oscillations at nuclear reactors, benefits from a very intense flux of $\bar{\nu}_e$ generated by β^- decays of the neutron-rich fission fragments in the reactor core. The huge intensity ($\approx 2 \times 10^{20}\bar{\nu}_e\,s^{-1}\,GW_{th}^{-1}$ for a pressurized water reactor) of this flux makes nuclear reactors one of the most attractive sources to perform systematic studies on neutrino properties[72]. As a matter of fact, since the neutrino discovery by Reines and Cowan [1] many experiments have been carried out at reactors, either to study the basic parameters of weak interactions or to look for physics beyond the standard model.

The neutrino flux from reactors has been extensively studied in the last few years, since the accuracy with which the reactor source is known , in the relevant energy regime, is often the factor that determines the quality of the test of the oscillation hypothesis. We shall show in this Chapter that the knowledge of the neutrino spectrum is now quite accurate (better than 2%), provided the reactor power is known at 1% (which is usually the case) and the fuel composition is also known (not a severe requirement).

3.1 Reactors as ν factories

Nuclear reactors generate thermal power by using the fission energy of nuclei in the core. The fuel is generally made of ^{238}U (the most abundant uranium isotope) and enriched with ^{235}U (which is by far the most fissile); a typical fission reaction involving a ^{235}U is the following:

$$n + {}^{235}U \rightarrow {}^{236}U \rightarrow {}^{140}Ba + {}^{94}Kr + 2n + kin.\ energy$$

$$
{}^{140}Ba \xrightarrow[\substack{13\,d \\ 1\,MeV}]{\beta^-} {}^{140}La \xrightarrow[\substack{40\,h \\ 2.2\,MeV}]{\beta^-} {}^{140}Ce \tag{3.1}
$$

$$
{}^{94}Kr \xrightarrow[\substack{0.2\,s \\ 7.5\,MeV}]{\beta^-} {}^{94}Rb \xrightarrow[\substack{2.7\,s \\ 10\,MeV}]{\beta^-} {}^{94}Sr \xrightarrow[\substack{75\,s \\ 3.4\,MeV}]{\beta^-} {}^{94}Y \xrightarrow[\substack{19\,m \\ 4.9\,MeV}]{\beta^-} {}^{94}Zr
$$

The fragmentation of a fissile nucleus generates two nuclei (whose charge may range from 30 to 65) plus two or three neutrons. The mass distribution of the resulting fission fragments has a typical "saddle" shape, as shown in Fig. 3.1; the most probable A values range from 86 to 104 or from 130 to 148. These fission products are usually much neutron-richer than their corresponding stable isotopes and tend to get rid of this excess by direct emission or, more frequently, by undergoing β^- decays. The sorting neutrons are then able to fission other nuclei, so giving rise to a chain reaction.

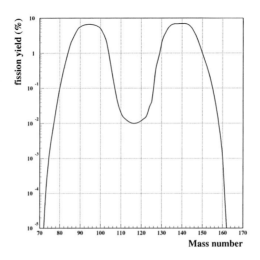

Figure 3.1: Mass distribution of the ^{235}U fission fragments.

Each β-decay is accompanied by the emission of an electron antineutrino. One can roughly estimate the $\bar{\nu}_e$ flux by knowing that the average number of β^- decays is about 6 per fission and the average thermal energy release amounts to about 203 MeV [73]. The neutrino emission rate is then related to the fission rate by

$$N_\nu \approx 6 \times N_f = 6\frac{W_{th}}{\langle E_f \rangle} \simeq \frac{6 \times 6.24 \times 10^{21}}{203} W(\mathrm{GW}_{th})\,\mathrm{s}^{-1}$$
$$= 1.8 \times 10^{20} W(\mathrm{GW}_{th})\,\mathrm{s}^{-1}, \tag{3.2}$$

W_{th} being the thermal power expressed in GW. The resulting flux is perfectly isotropic and with essentially no contamination from other neutrino flavours, since the ν_e emission rate (due to the β^+ decays or to electron captures by the neutron-poor isotopes) is lower than the previous one by a factor $10^{-5} \div 10^{-8}$

and may be neglected[1].

At the beginning of the reactor operation, the dominant contribution to the fission rate comes from ^{235}U. As a matter of fact, besides a lower fission cross section, ^{238}U nuclei have a threshold (~ 0.8 MeV) high enough to forbid any fission by thermal neutrons. On the contrary they may capture thermal neutrons thereby producing two fissile plutonium isotopes (^{239}Pu, ^{241}Pu); the first of these is generated through the reaction

$$n + {}^{238}\text{U} \longrightarrow {}^{239}\text{U} \xrightarrow{\beta^-} {}^{239}\text{Np} \xrightarrow{\beta^-} {}^{239}\text{Pu} \qquad (T_{1/2} = 24100\,\text{y}); \qquad (3.3)$$

the second isotope is produced by the following capture of two thermal neutrons.

The uranium and plutonium isotopes mentioned are the main components of the reactor fuel; the contribution from the other fissile nuclei to the thermal power is lower than 0.1%. The isotopic fuel composition significantly changes during the operating period of a reactor, as a consequence of the burning of the ^{235}U and the accumulation of plutonium isotopes in the reactor core. Since the $\bar{\nu}_e$ energy spectra following fission of the various isotopes differ from each other, an evaluation of the time-dependent composite antineutrino spectrum requires a detailed knowledge of the contribution to the reactor power of each fissile element, as a function of time.

In the following we shall describe how to follow the fuel evolution by using the information provided by the technical staff of the Chooz power station. We note that the adopted procedure is applicable to any PWR-type reactors, since they all have very similar operating modes.

3.2 The E.D.F. power plant at Chooz

3.2.1 Description and working

The Chooz nuclear power plant is located at the homonymous village in the north region of France, close to the border with Belgium, on the banks of the River Meuse. The power plant consists of two twin PWR reactors belonging to a generation (named N4) recently developed in France, whose main innovation consists in an improved power yield (4.25 GWth, 1.45 GWe at full operating condition), larger than any other PWR reactor.

The core of both reactors consists of an assembly of 205 fuel elements lying on a socket plate tied to the bottom of the reactor vessel. This vessel is filled with pressurized water ($p = 155$ bars) at a temperature ranging from 280 °C at the entrance to about 320 °C at the exit. The water (which

---

[1]In what follows, for the sake of brevity, we will often use the term "neutrino" instead of "antineutrino" to indicate the $\bar{\nu}_e$'s produced at reactors.

works at the same time as the neutron moderator and the cooling element
for the core) circulates by means of four loops[2], whose working principle
is illustrated in Fig. 3.2. Each loop involves a primary pump and a steam
generator made of 5416 tubes immersed in the water of a secondary loop at
a pressure (56 bars) lower than the one of the primary loop. As soon as the
primary water passes through these tubes, the secondary water is vaporized
and the steam produced goes to the turbine-alternator unit connected to the
electric system [74].

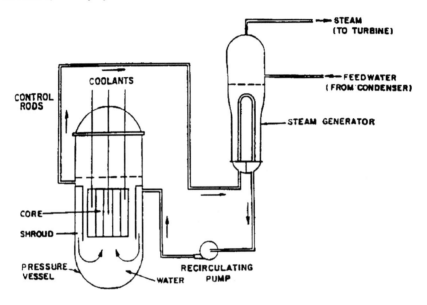

Figure 3.2: Working scheme of a PWR reactor.

The water in the primary loop is slightly doped by Boron which is a strong
absorber of thermal neutrons. As a matter of fact, the Boron activity gives
information on the trend of the chain reaction in the core. Another way to
keep the chain reaction under control is to use the so-called "control" rods,
which is a set of Boron-doped steel rods hanging from the top of the vessel.
The descent of the "control-cluster" causes the capture of neutrons and, as a
result, the slowing down of the chain reaction. Control rods are intended to
maintain the reactor in its critical state, for shutdown, for starting-up and
changing over from one power level to another[3].

---

[2]The name N4 ("nouveau 4 boucles") of the new PWR reactor generation is due the
number of water loops, which is for the first time larger than three.

[3]A crucial role in controlling the self-sustained chain reactions in nuclear reactors is
also played by the small amount ($\approx 0.7\%$) of delayed neutrons emitted in β−decay of some
fission fragments.

3.2.2 Reactor power monitor

Two methods were developed by the E.D.F. technical staff to monitor the total thermal power produced by each reactor. The first one is based on the heat balance of the steam generators. Measurements are made of the parameters (water flux and vapour tension in the secondary loop) needed to determine the amount of heat being exchanged at the steam generators. The resulting values are periodically available on a computer network dedicated to the purpose which records the data on an Excel file. The overall precision on the so-obtained thermal power is claimed to be 0.6%.

The second set of thermal power data is provided by the external neutron flux measurements. This flux is expected to be directly proportional to the fission rate inside the core and, as a consequence, to the released thermal power. For each reactor, four different neutron detectors (one proportional counters plus three ionization chambers) are located on the opposite sides of the reactor vessel. The precision of this method is poorer than the previous one, because of the spread in the energy release per fission due to the different fissile isotopes; the accuracy is estimated to be about 1.5%. However, this method has the advantage of operating permanently and is used at the same time as a power monitor and as a pilot of the reactor safety system. The neutron detector outputs are fed to the computer network and written twice a day (or more frequently if the power ramps up or down) to another Excel file. The computer is also programmed to drive a direct current generator whose amplitude is proportional to the neutron detection rate. In Chapter 4 we shall describe our on-line access to this current output and the format of the reactor power data packet. We shall also show how to use the files of the thermal measurements recorded by E.D.F. to calibrate the on-line data.

3.2.3 Map of the reactor core

The nuclear fuel consists of 110 T of uranium oxide tablets ($\phi = 8.2\,\text{mm}$) enriched with ^{235}U and piled in 4 m long, 1 cm wide assemblies; the standard enrichment for this type of reactor amounts to 3.1%. Each fuel element contains 264 assemblies. About 1/3 of the 205 fuel elements are changed at the end of each cycle. The remaining ones are displaced towards the centre and new fuel elements are arranged in the outer part of the core, so as to get the fuel burning as uniform as possible. For the start-up of the Chooz reactors, brand-new fuel rods were used. So, in order to reproduce the behaviour of reactors at equilibrium, where partially burned-up fuel is used, 137 less enriched elements are located at the centre (68 at 2.4% and 69 at 1.8%, while the standard enrichment is 3.1%). A schematic map of the reactor core is drawn in Fig. 3.3.

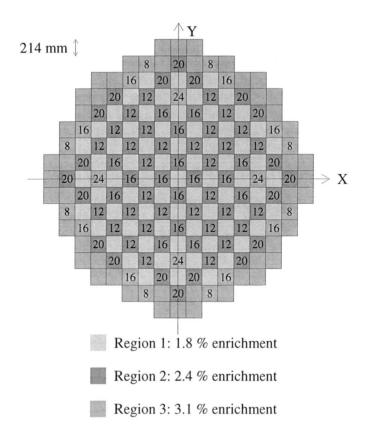

Figure 3.3: Schematic view of the fuel rods in the core for the first cycle of the Chooz reactors. The number of Boron poison rods assembled with each fuel element is also indicated.

As shown in Fig. 3.3, the fuel assemblies of the most enriched elements are surrounded by a number of 12.7% Boron-doped steel rods, termed "poison" rods[4]. Their number per fuel element varies according to the fuel loading as well as to the element position in the core; therefore this must be taken into account in computing the fuel evolution.

3.2.4   Fuel evolution

The unit used to describe the aging of the fuel at nuclear reactors is the MW d/T, which measures the amount of energy per ton extracted from the

---

[4]The significant feature of these poison rods is to absorb the thermal neutron excess, thus cumulating ^{11}B, which has negligible neutron absorption cross section, inside the core. Thus, as the fuel deplets, the poison burns out itself resulting in the compensation to the loss of the multiplication factor due to the burn-up.

nuclear fuel since its introduction into the reactor core. This quantity is called "burn-up" and is strictly related to the fissile isotope composition of the fuel.

For any PWR reactor, the procedure to compute the evolution of the fuel in the core needs information daily provided by the reactor technical staff. This includes:

- the daily cumulated burn-up, given as

$$\beta(t) \equiv \frac{1}{M_U} \int_0^t W_{th}(t') \, dt', \tag{3.4}$$

W being the thermal power, t the time since the start of reactor operation and $M_U = 110.26$ T the total amount of uranium in the core:

- the burn-up β_i and the relative contribution α_i to the power from the i-th fuel element, at several stages of the reactor cycle (as shown in Fig. 3.4):

- the fission rate percentage of the main fissile isotopes, as a function of the element burn-up, for any initial fuel loading and any number of poison rods.

The first two inputs determine the daily burn-up of each fuel element. These values are then used to obtain the relative power contribution f_k^i from the k-th fissile isotope for the i-th fuel element, according to a table provided by E.D.F. The number n_k^i of fissions per second for the i-th element for each isotope k can thereby be computed with the relationship

$$n_k^i(\beta) = \frac{\alpha_i(\beta) f_k^i(\beta) W(t)}{\sum_k f_k^i(\beta) E_k}, \tag{3.5}$$

where E_k is the energy release per fission of isotope k, whose values are listed in Tab. 3.1. Adding the contribution of all fuel elements yields the

isotope	energy (MeV)
^{235}U	201.7 ± 0.6
^{238}U	205.0 ± 0.9
^{239}Pu	210.0 ± 0.9
^{241}Pu	212.4 ± 1.0

Table 3.1: Energy release per fission of the main fissile isotopes (from ref. [75]).

average number N_k of fissions per second for each isotope k. The relative

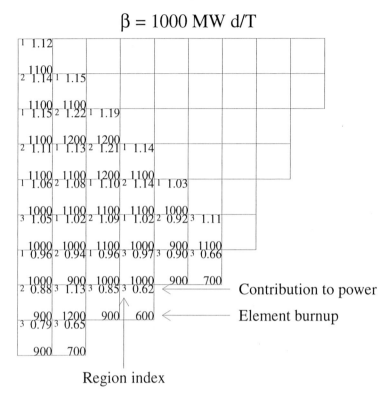

Figure 3.4: Power distribution and burn-up values for the fuel elements in an octant of the Chooz reactor core at a certain step ($\beta = 1000$ of the first cycle). The contribution to power of each element is normalized to have mean value equal to one.

contributions of the four relevant isotopes are shown in Fig. 3.5 for the first cycle of the Chooz reactors. Contributions from other fissioning isotopes, such as ^{236}U, ^{240}Pu, ^{242}Pu, etc. amount to less than 0.1% and are therefore neglected.

In order to obtain the source spectrum, we require, in addition to the average fission rate N_k of each of the four isotopes, the associated differential neutrino yields per fission $S_k(E_\nu)$. The data available for these yields are discussed in the next section.

3.3   The neutrino spectra

Two methods have been proposed to determine the electron antineutrino spectrum associated with a particular fissioning nucleus. The more straight-

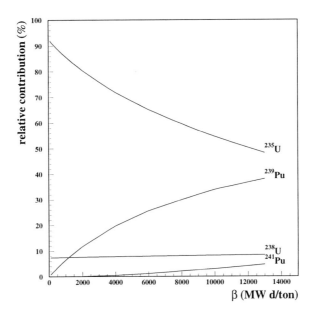

Figure 3.5: Contribution to the fission rate of the relevant fissile isotopes during the first cycle of the Chooz reactors.

forward one involves summation of all beta branches of all fission fragments. The second method (which is more trustworthy for reasons we will clarify below) relies on an experimentally determined electron spectrum associated with fission, which is then directly converted into the corresponding $\overline{\nu}_e$ spectrum. Let us briefly review both.

3.3.1 The "summation" approach

The summation method was first used by Perkins and King [76]. Later evaluations, with more complete experimental data, were reported by several authors [77, 78, 79, 80, 81]. The followed procedure, based on the knowledge of the fission yields, the branching ratios for the various allowed decays and their relative Q-values, goes as follows:

$$S(E_\nu) = \sum_{i,n} Y_n(Z, A, t) b_n^i (E_0^i) P(E_\nu, E_0^i, Z), \qquad (3.6)$$

where

- Y_n is the number of β^--decays per unit time of fission product (Z, A) after the fissioning material has been exposed to neutrons for a time [5]

---

[5] For t longer than the fission fragment lifetime this quantity converges to the cumulative fission yields and becomes independent of t.

t:

- $P(E_\nu)$ is the normalised, Coulomb corrected spectrum shape factor:

- b_n^i is the branching ratio for the decay of the n-th fission product into the i-th decay branch with end-point E_0^i.

While the first two points are well known, the most serious difficulty is related to the branching ratios b_n^i. The set of experimentally known branching ratios is rather limited. The number of possible fission fragments is extremely large (≈ 700) and only for about 250 of those, complete sets of experimental data are available; for all others, one has to make a reasonable guess. These latter fragments are generally neutron rich and have high Q-values; even though their contribution to the total fission yield is relatively small, they affect the high-energy tail of the $\overline{\nu}_e$ spectrum. As a result, the antineutrino spectra evaluated by several authors may differ significantly from each other in that energy range, even though the basic procedure is the same in all these works.

An important test of these predictions is given by the comparison with an experimentally determined electron spectrum associated with fission; as a matter of fact, the antineutrino and electron spectra are usually calculated together and similarly depend on input parameters and on theoretical assumptions; Fig. 3.6 summarizes the situation. In general, the evaluations do not reproduce the experimental spectra in a satisfactory way; the best agreement was achieved by Klapdor et al. [80], whose predicted spectrum differs from the experimental one by no more than 5% for each fissile isotope.

3.3.2   The "conversion" approach

A more reliable method to determine the $\overline{\nu}_e$ spectrum at reactors consists in measuring the electron spectrum emitted by a layer of fissile material activated by thermal neutrons and then converting the experimental electron spectrum into the $\overline{\nu}_e$ one. This procedure was first developed by Carter et al. [83] and tested by Davis et al. [84] and Borovoi [85]. Later and more precise measurements were reported by Schreckenbach et al. [82, 86, 87]. In the latter case thin foils enriched with the main fissile nuclei (about 1 mg) were exposed to the intense thermal neutron flux ($\approx 3 \cdot 10^{14}\,\mathrm{s}^{-1}$) coming out of the reactor core at the Institute Laue-Langevin in Grenoble. A high-resolution ($\delta p/p = 3.5 \cdot 10^{-4}$) β-spectrometer was used to measure the momentum of the emerging electrons.

The β^- spectrum for each fissile isotope was approximated by the superposition of a set of hypothetical allowed branches with amplitude a_i and

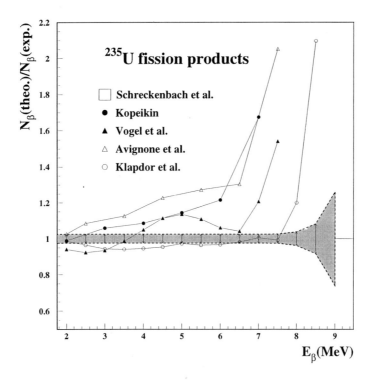

Figure 3.6: Comparison of calculated β^- spectra of ^{235}U fission products and the experimental result by Schreckenbach et al. [82]. The dashed band indicates the total uncertainties (at 90% C.L.) of the latter.

end-point E_0^i. The sum gives therefore

$$S_\beta(E_\beta) = \sum_i a_i S_\beta^i[E_\beta, E_0^i, \overline{Z}(E_0^i)], \qquad (3.7)$$

where S_β^i is the spectrum shape of the i-th branch and the summation is over the branches with end-points larger than E_β. The spectrum weakly depends also on the charge $\overline{Z}$ (averaged over the β-decaying nuclei with end-point E_0^i) because of the Coulomb interaction in the final state. The measured electron spectrum was then used to determine the set of values $\{a_i, E_0^i\}$ by means of a fit procedure. The (3.7), with the introduction of the best-fit parameters, reproduces the measured spectrum to better than 1%.

The β^- spectrum for each individual, hypothetical branch was then converted to the correlated $\overline{\nu}_e$ spectrum under the assumption that both the electron and the antineutrino share the total available energy E_0^i. Thus, for each branch with end-point E_0^i, the probability of emitting an electron with energy E_β is equal to the probability of having a $\overline{\nu}_e$ of energy $E_0^i - E_\beta$. From

(3.7), by introducing the fit parameters, one obtains

$$S_\nu(E_\nu) = \sum_i a_i S_\beta^i[(E_0^i - E_\nu), E_0^i, \overline{Z}(E_0^i)] \tag{3.8}$$

These yields contain a normalization error of 1.9% originating from the error
on the neutron flux and from the absolute calibration uncertainty of the spectrometer. The conversion procedure introduces a global shape uncertainty
into the neutrino spectrum beyond the inherent experimental errors. The
main sources of this additional uncertainty, ranging from 1.34% at 3 MeV
and 9.2% at 8 MeV, are the scattering in the nuclear charge distribution and
the higher-order corrections (higher Coulomb terms and weak magnetism, for
which an uncertainty of the order of the correction term itself was assumed).

Such a method was adopted to obtain the neutrino yields of the ^{235}U,
^{239}Pu and ^{241}Pu fissions; the resulting spectra are presented in Fig. 3.7.
Unfortunately, no experimental data is available for ^{238}U which cannot be
fissioned by thermal neutrons. So we are forced to trust the theoretical predictions by [79] in order to estimate the contribution to the $\overline{\nu}_e$ spectrum by
the ^{238}U fission products. Although such predictions are less reliable than a
direct measurement, one has to note that the contribution to the number of
fissions, due to this isotope, is quite stable and never larger than 8%, so that
a possible discrepancy between the predicted and the real spectrum should
not lead to significant errors.

3.3.3   Systematic uncertainties of the neutrino spectrum

In past experiments at reactors, e.g. at Gösgen and Bugey, more than $3 \times$
10^4 neutrino events were recorded at each position. Since no evidence for
oscillation was observed, these experiments can be interpreted as a check
of a reactor as a neutrino source of known characteristics at a few percent
level. The statistical accuracy reached by these experiments is higher than
the discrepancies among the proposed reactor neutrino spectra; it is then
possible to use these results to discriminate between the existing models.

The Bugey 3 collaboration [88] measured the positron energy spectrum at
15 and 40 m from the reactor core and compared their data with the results
of a Monte Carlo simulation using the neutrino spectrum models proposed
by [86, 80, 81]. As one can see in Fig. 3.8, the data are in excellent agreement
with the measurements made at ILL, while the agreement with the other
models is poorer.

An improved measurement of the integral neutrino flux was performed in
1993. The measurement used an integral type detector previously employed

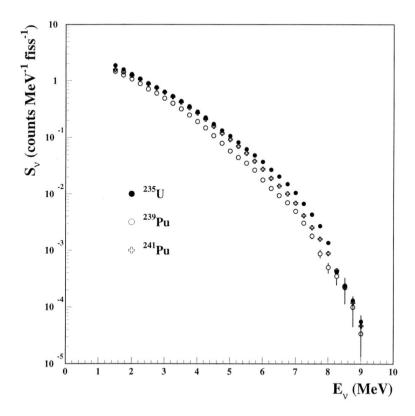

Figure 3.7: Neutrino yield per fission of the listed isotopes, as determined by converting the measured β spectra [86, 87].

at the Rovno nuclear power plant [73] and subsequently moved to Bugey [89]. In that detector only neutrons were detected in ^{3}He counters; positrons were not detected. The apparatus was installed at 15 m from the reactor core (in the same area where the Bugey 5 experiment was carried out). About 3×10^5 neutrino events were collected so that the reaction rate was determined with 0.67% statistical accuracy. The neutrino event rate corresponds to a certain average fuel composition and is related to the cross section per fission σ_f by

$$n_\nu = \frac{1}{4\pi R^2} \frac{W_{th}}{\langle E_f \rangle} N_p \varepsilon \sigma_f \tag{3.9}$$

The fuel composition, the thermal power W, the E_f absorbed in the reactor core per fission and the distance R were provided by the EdF–Bugey technical staff. The efficiency of the neutron detection ε was carefully measured; the overall accuracy was estimated to be of 1.4%. The experimental result was then compared to the expected neutrino flux, which can be inferred by

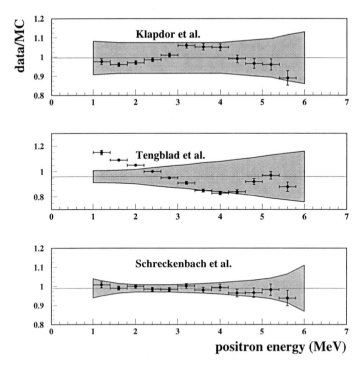

Figure 3.8: Comparison of Bugey 3 data with three different reactor spectrum models. The error bars include only the statistical uncertainties. The dashed lines are the quadratic sum of the quoted error of the models and the error due to the energy calibration.

introducing the neutrino spectra S_k obtained at ILL[82, 86, 87], the cross section of the detection reaction (2.22) and the reactor parameters in (3.9). The expected cross section per fission for reaction (2.22) is given by

$$\sigma_f = \int_0^\infty \sigma(E_\nu) S(E_\nu)\, \mathrm{d}E_\nu = \sum_k f_k \int_0^\infty \sigma(E_\nu) S_k(E_\nu)\, \mathrm{d}E_\nu = \sum_k f_k \sigma_k$$

(3.10)

where f_k are the contribution of the main fissile nuclei to the total number of fission, S_k their corresponding $\bar\nu_e$ spectrum and σ_k their cross section per fission for that reaction. They measured $\sigma'_f = 5.752 \times 10^{-19}$barns/fission$\pm 1.4\%$, in excellent agreement with the expected value ($\sigma_f = 5.824 \times 10^{-19}$barns/fission$\pm 2.7\%$) and twice more accurate than the predictions based on the knowledge of the neutrino spectrum.

We are thus led to adopt the ILL spectra measurements to reproduce the neutrino spectral shape; however, these spectra still need to be renormalized

in order to fit the integral neutrino flux determined at Bugey 5. The reaction
cross section (3.10) can be rewritten in the form

$$\sigma_f = \sigma_f' + \sum_{k=1}^{4}(f_k - f_k')\sigma_k = \sigma_f' + \sum_{k=2}^{4}(f_k - f_k')(\sigma_k - \sigma_1) \qquad (3.11)$$

where f_k' were the contributions to the total amount of fissions at the Bugey
reactors during the above measurements (in the last equality we used $\sum_k f_k = \sum_k f_k' = 1$). In Fig. 3.9a the combined (ILL+Bugey) cross section (3.11)
obtained for the Chooz reactors is divided by the ILL cross section (3.10)
and plotted vs. the reactor burn-up; the average ratio then amounts to
0.987. By combining the uncertainty on the neutrino spectra, on the cross
section for the reaction (2.22) and on the fission contributions f_k (which are
of the order of 5%), we obtained the relative error on the neutrino detection
rate, as a function of the fuel burn-up. As shown in Fig. 3.9b, the average
error on the first reactor cycle is lowered from 2.4% (ILL data alone) to
1.6% (ILL+Bugey 5). Other minor sources of errors come from the residual
neutrino emission from long-lived fission fragments (which we will deal with
in the next Section); however, one can consider the reactor source as known
at the 2% level. Therefore the neutrino oscillation search at reactors does
not any longer need to monitor the neutrino flux at a close position with a
separate detector.

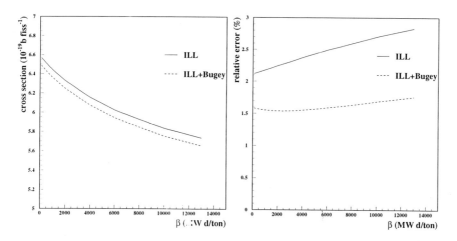

Figure 3.9: Comparison of the combined (ILL+Bugey) reaction cross section with
the ILL cross section (left) and their relative error (right) as a function of the first
Chooz reactor cycle burn-up.

3.3.4   Neutrino spectrum time relaxation and residual neutrino emission

Another possible source of uncertainty of the neutrino flux is related to the residual emission due to the β^- decay of long-lived fission fragments. If one takes into account this further contribution, the linear relation (3.9) between the prompt $\bar{\nu}_e$ interaction rate and the current thermal power W no longer holds. Nevertheless, the $\bar{\nu}_e$ spectra determined at ILL were derived after about 1.5 d exposure time, so that neutrinos from longer-lived fragment decays are not included. The expected neutrino rate based on this model may thus be underestimated with respect to the experimental data. Fortunately, only in a few cases the maximum neutrino energy is above the reaction (2.22) threshold since, as one would expect, the longer the lifetime, the lower the Q-value of the decay[6]. This effect has been evaluated by using the cumulative yields of the known long-lived fission fragments [90]; the results for ^{235}U and ^{239}Pu are summarized in Tab. 3.2. In particular, the reaction cross section σ_f

Table 3.2: Time evolution of neutrino spectra relative to infinite irradiation time (from [90]).

E_ν (MeV)	^{235}U			^{239}Pu		
	10^4 s	1.5 d	10^7	10^4 s	1.5 d	10^7
1.5	0.837	0.946	0.988	0.861	0.949	0.990
2	0.897	0.976	0.992	0.904	0.968	0.986
2.5	0.925	0.981	0.990	0.939	0.975	0.986
3	0.963	0.997	1.000	0.967	0.989	0.993
3.5	0.967	1.000	1.000	0.979	0.997	1.000

computed by using (3.10) is probably by $\approx 0.3\%$ lower than the effective one. This systematic shift affects the accuracy on the reaction cross section; we will then assume an overall 1.9% uncertainty on the integral neutrino rate.

.

3.4   Neutrino detection

3.4.1   The inverse beta-decay reaction

The detection of reactor antineutrinos is usually based on the reaction (2.22). This is the most suitable process since:

---

[6]This point might be much more relevant in the case of experiments looking for neutrino elastic scattering interactions (i.e., measurement of the neutrino magnetic moment) and needs more careful treatment.

- it has the highest cross section in the energy range of reactor antineutrinos (see table 3.3):

- provides a convenient time correlated pair of positron and neutron signals, which allows us to cut down most of the background.

The antineutrino and the positron energy are related by

$$E_{\bar{\nu}_e} = E_{e+} + (M_n - M_p) + \mathcal{O}(E_{\bar{\nu}_e}/M_n), \tag{3.12}$$

where the infinitesimal term corresponds to the neutron recoil. Thus a measurement of the positron energy allows an accurate determination of the energy of the incoming antineutrino. The threshold for the reaction (2.22) is 1.804 MeV, equal to the nucleon mass difference plus the positron mass. In the low energy limit, the cross section for the reaction (2.22) may be written as a function of outgoing positron energy as follows:

$$\sigma(E_{e+}) = \frac{2\pi^2\hbar^3}{m_e^5 f \tau_n} p_{e+} E_{e+} (1 + \delta_{rad} + \delta_{WM}) \tag{3.13}$$

The transition matrix element has been expressed in terms of the free neutron decay phase-space factor $f = 1.71465(15)$ [91] and lifetime $\tau_n = (886.7 \pm 1.9)s$ [92]. Two higher-order correction terms (both of 1% order of magnitude) are also included:

(i) a radiative correction of the order of α, including a contribution of internal bremsstrahlung, which can be approximated by

$$\delta_{rad}(E_{e+}) = 11.7 \times 10^{-3}(E_{e+} - m_e)^{-0.3} \tag{3.14}$$

with the positron energy expressed in MeV [93]:

(ii) a correction for weak magnetism, arising from the difference $\mu = \mu_n - \mu_p = -4.705890(2)\mu_N$ between the anomalous magnetic moment of the neutron and the proton

$$\delta_{WM}(E_{e+}) = -2\frac{\mu\lambda}{1 + 3\lambda^2}(E_{e+} + \Delta\beta p_{e+})/M_p, \tag{3.15}$$

where $\lambda = g_A/g_V = 1.2601 \pm 0.0025$ is the ratio of axial-vector and vector coupling constants and Δ is the nucleon mass difference [94].

The knowledge of the cross section is then much more accurate than the $\bar{\nu}_e$ spectrum, the major limitation being related to the uncertainty on τ_n (whose relative error is anyway less than 0.3%).

3.4.2   Other possible targets for low energy neutrinos

The table (3.3) summarizes the low energy $\bar{\nu}_e$-induced reactions; the listed σ_f values are the reaction cross sections folded with the $\bar{\nu}_e$ energy spectrum for ^{235}U; for electron scattering an effective threshold of 0.5 MeV for the electron recoils was assumed. Reactor antineutrinos can scatter on electrons by

Table 3.3: Reactor neutrino induced reactions

reaction	threshold (MeV)	$\sigma_f(10^{-44}\,\mathrm{cm^2/fission})$
$\bar{\nu}_e + \mathrm{p} \rightarrow \mathrm{e}^+ + \mathrm{n}$	1.8	63.4
$\bar{\nu}_e + \mathrm{e}^- \rightarrow \bar{\nu}_e + \mathrm{e}^-$	0.7	1.0
$\bar{\nu}_e + \mathrm{d} \rightarrow \bar{\nu}_e + \mathrm{p} + \mathrm{n}$	2.2	3.2
$\bar{\nu}_e + \mathrm{d} \rightarrow \mathrm{e}^+ + \mathrm{n} + \mathrm{n}$	4.0	1.1
$\bar{\nu}_e + {}^{12}\mathrm{C} \rightarrow \mathrm{e}^+ + {}^{12}\mathrm{B}$	14.4	0
$\bar{\nu}_e + {}^{16}\mathrm{O} \rightarrow \mathrm{e}^+ + {}^{16}\mathrm{N}$	11.4	0

a combination of the charged and neutral current weak interactions. Besides its cross section (about a factor 50 lower than reaction (2.22)), it lacks a double signature allowing a simpler discrimination from the background. This process would acquire much more interest if the electron antineutrinos had a non-vanishing magnetic moment; in such a case they could electromagnetically scatter on electrons with an amplitude proportional to μ_ν and going as $1/T$ when $T \rightarrow 0$ (T is the recoil electron kinetic energy). This suggests that reactors (which copiously produce low energy neutrinos) are a good place to look for the neutrino magnetic moment.

The charged current interaction processes involving nuclei heavier than protons have higher thresholds. In the case of ^{12}C and ^{16}O the whole neutrino spectrum is under threshold (see table 3.3), so that they are of no use at reactors. Even the reactions involving the deuterium have much smaller cross sections (when compared to (2.22)) due to the higher threshold; this makes the observation of these reactions quite difficult. Nevertheless these reactions were used by Reines et al. [95] to look for neutrino oscillations. The point is that oscillations $\bar{\nu}_e \rightarrow \bar{\nu}_x$ would cause a decrease in the rate of charged-current interactions, while the rate of a neutral-current reaction, being independent of the neutrino flavour, would not be affected. Therefore, the ratio of charged over neutral current reaction yields provides a test of neutrino oscillations almost independent of the knowledge of neutrino reactor spectrum.

3.4.3   Detection techniques

Neutrinos identified through reaction (2.22) can be detected by measuring the emerging, time correlated positrons and neutrons. The positron signal arises from the ionization loss plus the two annihilation 511 KeV quanta. The neutron (whose kinetic energy is less than 50 KeV) is moderated by elastic collisions on the target nuclei down to the thermal energy. Thus a neutrino detector has to fulfil two general requirements:

(1) it must be made of hydrogen-rich materials, since protons are at the same time the most suitable neutrino target and the most efficient neutron moderator:

(2) it must contain nuclei with a high thermal neutron capture cross section to obtain the characteristic neutron signal (a list of the most interesting nuclei is presented in table 3.4).

Table 3.4: List of thermal neutron capture reactions

nucleus	σ (barns)	signature
p	0.3	$n + p \rightarrow d + \gamma(2.2\,\text{MeV})$
Cd	2450	$n + Cd \rightarrow Cd^* \rightarrow Cd + \sum_i \gamma_i(9\,\text{MeV})$
Gd	50000	$n + Gd \rightarrow Gd^* \rightarrow Gd + \sum_i \gamma_i(8\,\text{MeV})$
^{10}B	767	$n + {}^{10}B \rightarrow \alpha + {}^7Li,\ Q = 2.8\,\text{MeV}$
^3He	5350	$n + {}^3He \rightarrow p + {}^3H,\ Q = 765\,\text{KeV}$
^6Li	940	$n + {}^6Li \rightarrow {}^4He + {}^3H,\ Q = 4.8\,\text{MeV}$

A neutrino oscillation test based on the analysis of the spectral shape requires a differential detector to measure the positron energy deposit. In homogeneous detectors, the target is also the positron and neutron detector; they are generally made of liquid scintillator doped by one of the nuclei listed in table 3.4; the amount of neutron-capturing nuclei is chosen to maximize the neutron efficiency detection and shorten as much as possible the capture time. In segmented detectors the neutron detection does not take place in the target, but in specific detectors (typically ^{3}He proportional tubes); in this case the geometry of the detector is so chosen as to get the highest neutron efficiency, while the neutron capture time is normally longer than in the previous case ($\approx 200\,\mu s$ against $10 \div 50\,\mu s$).

The positron detection is not mandatory for a measurement of the integral neutrino flux (just like the one at Rovno). One can use a passive neutrino target (water for instance) within an apparatus optimized to detect neutrons from (2.22).

3.5   Reactor simulation

The neutrino source spectrum is calculated by means of a Monte Carlo simulation of the reactor. Let us review in detail how this simulation works.

Neutrino production date

The neutrino production rate, apart from correction due to burn-up effects, is proportional to the reactor thermal power. So, in order to evaluate the daily number of neutrino interactions in our detector, we define a neutrino integrated luminosity as the daily average thermal power $\overline{W}(k)$ multiplied by the detector lifetime $T_l(k)$ on the k-th day. This quantity is used to draw lots for the date of the neutrino production. For a certain data taking period from day d_1 to day d_n, we have to build the cumulative distribution function

$$F(d_j) = \frac{\sum_{k=1}^{j} \overline{W}(k)T_l(k)}{\sum_{k=1}^{n} \overline{W}(k)T_l(k)}, \qquad (j = 1, \ldots, n) \qquad (3.16)$$

Then, after choosing u from a random distribution on $(0, 1)$, we find d_j such that

$$F(d_{j-1}) < u \le F(d_j)$$

Burn-up and thermal power distribution

Let β be the cumulated core burn-up on the extracted date and β^s the burn-up step value such that $\beta^{s-1} < \beta \le \beta^s$. It is then straightforward to extract the piece of information concerning the fuel elements (power contribution and burn-up) from the data placed at our disposal by the reactor technical staff. For instance, the fuel element i contributes to the total power according to the simple formula

$$\alpha_i(\beta) = \frac{\alpha_i^s - \alpha_i^{s-1}}{\beta^s - \beta^{s-1}}(\beta - \beta^{s-1}) + \alpha_i^{s-1}, \qquad (3.17)$$

where $\alpha_i^s = \alpha_i(\beta^s)$ are the values to be extracted from the table provided by E.D.F. An identical procedure is applied to determine the burn-up β_i of the i-th fuel element, once the tabulated values β_i^s are given.

The power distribution function can be used to generate the neutrino production point. According to the inverse transform method, it is possible to define a cumulative power distribution function and extract a fuel element j such that

$$\sum_{i=1}^{j-1} \alpha_i < u \leq \sum_{i=1}^{j} \alpha_i,$$

$u \in (0, 1)$ being a randomly generated number. The coordinates of the element j, in a reference frame with the origin located at the core centre and with the X, Y axis directed as shown in Fig. 3.3, are expressed as $\vec{X_j} = (m, n, 0)\Delta$, where $\Delta = 214\,\text{mm}$ is the transverse step size of an element rod. The neutrino production point is randomly generated within the extracted element; if $\vec{x_f}$ are the neutrino coordinates in the element frame, then in the reactor frame these coordinates become[7]

$$\vec{X_\nu} = \vec{X_j} + \vec{x_f} = (m\Delta + x_f, n\Delta + y_f, z_f), \tag{3.18}$$

Fuel composition and neutrino spectra

In order to obtain the average $\bar{\nu}_e$ energy spectrum per fission, we still need the relative contribution from the four main fissile isotopes to the total fission rate, for the fuel element where the neutrino is generated and for the extracted date. From the tables at our disposal, we know the coefficients f_k^i at each burn-up step of the first cycle. The behaviour of this quantity has been parametrized by fitting the tabulated values by means of the following function:

$$f_k^i(\beta^s) = p_{0k}^i + p_{1k}^i \cdot \exp(p_{2k}^i \beta^s) \tag{3.19}$$

Different parameter sets are determined, depending both on the initial fuel loading and on the number of poison rods for the considered element.

Finally, let us use the neutrino spectra $S_k(E_\nu)$ to weight the neutrino energy; this is generated according to a uniform distribution in the range $E_T \leq E_\nu \leq 10\,\text{MeV}$ (E_T is the energy threshold for the reaction (2.22)).

---

[7]Unfortunately we have no information yet about the power distribution along the reactor axis. We are then forced to assume a uniform distribution both in power contribution and burn-up. This might be a raw approximation, since the burn-up is expected to vary along the length of a fuel element because the neutron flux is smaller at the two ends. Furthermore, this distribution is not guaranteed to be symmetric, because of the use of the control rods. Nevertheless the longitudinal size of the reactor is much smaller than the average distance from the detector, so that any displacement of the power core barycentre would be unimportant.

Then, according to the (3.5), the composite neutrino spectrum results from

$$S_\nu(E_\nu) = \frac{\sum\limits_{k=1}^{4} f_k^i(\beta) S_k(E_\nu)}{\sum\limits_{k} f_k^i(\beta) E_k} \overline{WT_l},\qquad(3.20)$$

where $\overline{WT_l}$ is the daily neutrino luminosity averaged over the data taking period. For instance, Fig. 3.10 shows the neutrino spectrum obtained at Chooz on the start-up day ($\beta = 0$) and at an intermediate step ($\beta = 7000$) of the first reactor cycle. Due to the decrease of the ^{235}U concentration, a reduction of the neutrino interaction rate is observed and a softening of the neutrino spectrum is expected.

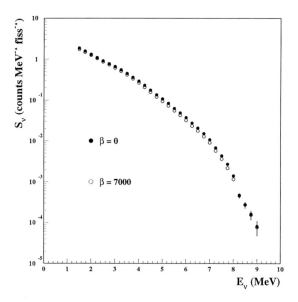

Figure 3.10: Comparison between the neutrino spectrum at the beginning and during the first cycle of the Chooz reactors.

Cross section and positron spectrum

The energy spectrum of the positrons coming from (2.22) is essentially the antineutrino spectrum shifted in energy and weighted by the cross section

(3.13). So, following (3.9), each positron is assigned a weight given by

$$S_{e^+}(T_{e^+}) = \frac{N_p}{4\pi d^2}\sigma(E_\nu)S_\nu(E_\nu), \tag{3.21}$$

where d is the distance from the neutrino production point to the detector and the positron kinetic energy T_{e^+} is given by (3.12). Fig. 3.11 shows the positron yield obtained by generating the neutrino spectra drawn in Fig. 3.10 in one day of data taking with both reactors at full power. Although the generated neutrino luminosity is the same, the decrease of the positron yield with the reactor operating time is evident. The evolution of the positron spectrum must be followed very accurately in order to reproduce the hardware threshold effects on the positron detection.

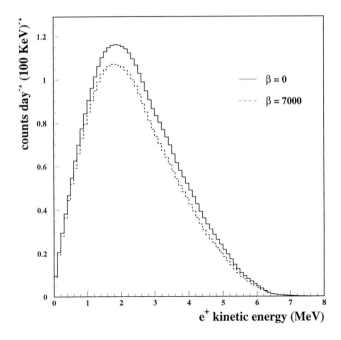

Figure 3.11: Positron spectra at the startup and during the first cycle of the Chooz reactors at maximum daily neutrino luminosity.

As an immediate consequence, also the integral neutrino interaction rate is expected to vary significantly during the reactor fuel cycle. A decrease of about 10% is foreseen for the cross section per fission (which is linear with the interaction rate according to (3.9)) during the first cycle of the Chooz reactors, as shown in Fig. 3.12. In Chapter 7 the measured neutrino rate

as a function of the burn-up will be shown and compared to the expected
behaviour, under the no-oscillation hypothesis.

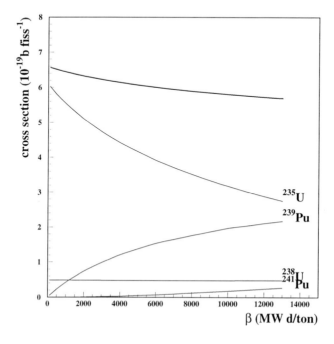

Figure 3.12: Cross section per fission as a function of the reactor burn-up. The
contribution of each fissile isotope is also shown.

Chapter 4

The neutrino detector

4.1 The site

The experimental hall is located in an underground tunnel at a 115 m (300 MWE) depth which had served as a service tunnel for the old underground Chooz power station. The site is located on the opposite side of the River Meuse with respect to the reactors, at an average distance of about 1 km from them. The thick rock overburden provides a strong suppression of the cosmic ray flux by a factor ~ 300, thus attenuating the cosmic-muon induced background. This is one of the main quality factors of the experiment; as shown in Fig. 4.1, the rock shielding preserves the signal to noise ratio of previous reactor experiments, in spite of a reduction by a factor ~ 100 of the neutrino flux due to the longer distance from the reactors.

4.2 The detector

4.2.1 Design and working principle

The neutrino detector consists of a vertical cylindrical steel tank located in a hole excavated in the rock of the gallery. The main tank contains, as shown in Fig. 4.2, three concentric egg-shaped volumes:

1) an inner volume (also indicated as Region I), working as the neutrino target:

2) an intermediate volume (Region II), working as an energy containment buffer:

3) an outer volume (Region III), working as an active "veto" for through cosmic muons.

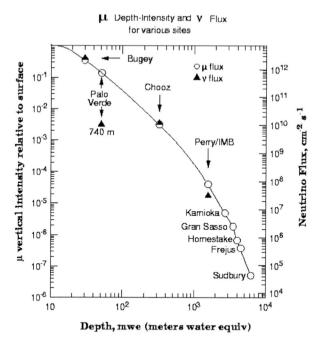

Figure 4.1: Cosmic muon flux compared to the neutrino flux at the different underground experimental sites. In the Chooz case the lower neutrino flux is compensated by the reduction of the muon flux.

The apparatus is conceived as a liquid scintillator low energy, high-resolution calorimeter. The detector geometry (a central volume of scintillator surrounded by photomultipliers) is common to the Borexino, LSND and SNO detectors.

The target region contains a Gd-doped liquid scintillator developed on purpose to detect neutrons. As a matter of fact, the neutrino detection is based on the delayed coincidence between the prompt positron signal generated by reaction (2.22) and boosted by the annihilation γ-rays and the signal associated with the γ-ray emission following the neutron capture reaction

$$ n + Gd \rightarrow Gd^{\star} \rightarrow Gd + \sum_i \gamma_i \qquad (4.1) $$

The choice of a Gd-doping is to maximize the neutron capture efficiency; as shown in Tab. 3.4, gadolinium has the highest thermal neutron cross section. Moreover, the large total γ-ray energy ($\approx 8\,\text{MeV}$, as shown in Tab. 4.1) allows for a much easier identification of the signals due to neutron capture from those associated with the natural radioactivity (whose energy does not exceed $3.5\,\text{MeV}$).

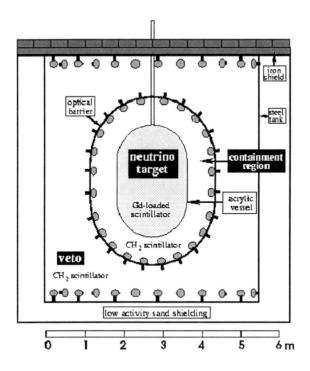

Figure 4.2: Layout of the Chooz detector.

Table 4.1: Abundances and thermal neutron cross sections for the Gd isotopes.

Gd isotope	$\sum_i E_{\gamma i}$ (KeV)	Abundance (%)	Cross section (barns)	Relative intensity
152	6247	0.20	735	$3 \cdot 10^{-5}$
154	6438	2.18	85	$3.8 \cdot 10^{-5}$
155	8536	14.80	60900	0.1848
156	6360	20.47	1.50	$6 \cdot 10^{-6}$
157	7937	15.65	254000	0.8151
158	5942	24.84	2.20	$1.1 \cdot 10^{-5}$
160	5635	21.86	0.77	$3 \cdot 10^{-6}$

Region II is filled with an unloaded high-flash point liquid scintillator. It provides a high-efficiency containment of the e.m. energy deposit; the efficiency for positron generated by neutrino interaction in the target is higher than 99% for $E \geq 1\,\mathrm{MeV}$; the containment of the γ-rays due to the neutron capture on Gd is slightly lower than 95% for an energy deposit $E > 6\,\mathrm{MeV}$.

The intermediate volume is bound by the "geode", an opaque plastic structure working as a support for the 192 inward-looking photomultiplier tubes (PMT from now on).

The outer volume, also filled with the unloaded scintillator of Region II, is the so-called "Veto" region. An additional 48 PMT's, arranged in two circular rings located at the top and the bottom of the main tank, detects the scintillation light yield associated with through-going cosmic muons. The Veto signal is used to tag and reject this major source of background. The outer scintillator layer is also wide enough to shield the neutrino target against the natural radioactivity from the surrounding materials.

4.2.2   The target

The inner detector volume is separated from Region II by a transparent 8 mm-thick vessel (the "ampoule"), shaped as a vertical cylindrical surface joined to two hemispherical end-caps. The outer radius of the cylinder and the end-caps is 90 cm, the height of the cylinder is 100 cm; the inner volume is 5.555 m^3, while the mass is 150 kg (empty). The vessel is made of an acrylic polymer (Altuglass), chosen for its excellent optical and mechanical properties and for its chemical resistance to scintillator aromatic compounds. The upper part of the vessel is fitted with a flange ($\phi = 70$ mm) to allow passage of filling pipes, calibration sources and temperature and pressure sensors.

4.2.3   The geode

The geode has the same shape as that of the target vessel, but with larger size; the cylinder height is the same, whereas the outer radius is 160 cm. The inner volume, subtracted by the target volume, is 19.6 m^3. The geode surface (a drawing of which is shown in Fig. 4.3) has a total area of 42 m^2 segmented into 32 panels; each panel is equipped with 6 8″ PMT's detecting the scintillation light produced in Regions I and II. The global PMT coverage is then 15%. In opposition to the acrylic inner vessel, the geode surface is required to be opaque so as to optically shield the inner Regions from the scintillation light produced in Region III. The external surface is white-coated in order to enhance the light collection in Region III and improve the Veto rejection efficiency; the inner surface is black-painted to reduce the light reflections, which could deteriorate the localization of the energy deposit in the detector.

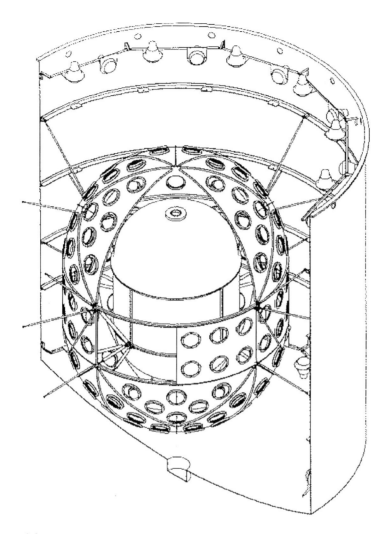

Figure 4.3: Mechanical drawing of the detector; the visible holes on the geode are for the PMT housing (from [96]).

4.2.4   The main tank

The main tank is a vertical steel cylinder with 5.5 m diameter and 6.2 m height, with two end-caps with a 7 m radius of curvature. The inner volume is 141 m^3 and the empty weight is 9.9 tons. The hollow space ($\approx$ 75 cm wide) between the tank and the concrete is filled with a low-radioactivity sand in order to improve the shielding against the external natural radioactivity.

4.3   The scintillators

About 5 tons of Gd-loaded scintillator and 107 of unloaded scintillator were needed for the experiment. Optimized formulations for both types were developed after testing at Drexel University and the University of New Mexico. The main properties of the two scintillators are listed in Tab. 4.2.

Table 4.2: Main properties of the liquid scintillators used in the experiment.

	Gd-loaded	unloaded
Chemical content:		
basic	Norpar-15 (50% vol.)	Mineral oil (92.8% vol.)
aromatics, alcohols	IPB+hexanol (50% vol.)	IPB (7.2% vol.)
wavelength shifters	p-PTP+bis-MSB (1 g/l)	PPO + DPA (1.5 g/l)
Atomic mass composition:		
H	12.2%	13.3%
C	84.4%	85.5%
Gd	0.1%	
others	3.3%	1.2%
compatibility	acrylic, teflon	
density (20 °C)	0.846 g/l	0.854 g/l
Flash point	69 °C	110 °C
Scintillation yield	5300 photons/MeV ($\simeq$ 35% of anthracene)	
Optical attenuation length	4 m	10 m
Refractive index	1.472	1.476
Neutron capture time	30.5 μs	180 μs
Thermal neutron capture path length	$\sim$ 6 cm	$\sim$ 40 cm
Capture fraction on Gd	84.1%	

The solution of the gadolinium salt $Gd(NO_3)_3$ in hexanol as well as the mixing of the basic and aromatic compounds was performed in a dedicated hall (*salle de mélange*) close to the gate of the underground tunnel. The amount of gadolinium (0.1 % in weight) was chosen in order to optimize the neutron capture time and the neutron detection efficiency; the measured values for the average capture time, path length and capture efficiency are listed in Tab. 4.2. A higher concentration would require more alcohol, which could lower the high flash point of the solution below the limit imposed by the safety rules in force in the E.d.F. power plants. Moreover, as we shall see later, the presence of the nitrate ions in solution progressively deteriorates the optical properties of the scintillator; therefore a higher concentration would further compromise the chemical stability of the scintillator.

A fundamental parameter for normalization of the neutrino event rate is the number of free protons in the target. An accurate evaluation of this number relies on precise measurements of the density and the hydrogen content of the Gd-loaded scintillator. The hydrogen content was determined by combustion of scintillator samples; the presence of volatile elements (alcohols first of all) makes this measurement particularly difficult. The value listed in Tab. 4.2 results from averaging the dozen measurements performed on scintillator samples at different laboratories [97, 98]; the overall relative uncertainty on the number of protons is 0.8%.

4.3.1 Study of the optical properties

The scintillator transparency showed a degradation over time, resulting in a slight decrease of the photoelectron yield; the most probable cause was singled out in the reduction of the nitrate ion (known to be an oxidative agent). On one side, this forced us to repeatedly check the attenuation length in the detector by using calibration sources all along the acquisition period; the followed procedure will be described in §6.4. In parallel, the Pisa group decided to start a systematic study of the chemical and optical stability of liquid Gd-loaded scintillators, with possible developments for future low-energy neutrino detectors[1]. A chemical laboratory was equipped with new instrumentation (glassware, benches, vacuum/Nitrogen lines, balance, safety cabinets, for preparing, proper handling and storing chemicals and scintillator samples).

The main optical properties were determined by means of both spectrophotometric and fluorimetric measurements. A Varian Cary 2200, UV-vis, 10 cm cell double-beam spectrophotometer was used to measure the absorbance of scintillators as a function of the beam light wavelength in the $350 \div 600$ nm range, the absorbance being defined as

$$A \equiv -\log_{10} \frac{I}{I_0} \qquad (4.2)$$

where I_0, I are the intensity respectively of the incident and the emerging beams. Several paired Hellma cells, with Suprasil quartz windows were used

---

[1]Several proposed experiments for solar neutrino detection aim at developing stable organic scintillators or crystals doped by metallic or Lanthanid species (such as Gd, Yb) for low energy neutrino spectroscopy [99]

for such measurements. The measurement of attenuation lengths of a few meters by means of 10 cm cells required great care in the stability control of the apparatus as well as an accurate cleaning of the cell optical surfaces. Corrections were applied to take into account the multiple reflections inside the cell due to the different refraction indices of air, quartz and scintillator; the overall systematic uncertainty on the absorbance (including the spectrometer stability and the planarity of the quartz windows) resulted to be 0.5%. The measured absorbance values were then converted into the attenuation length according to the following expression

$$\Lambda = 2.30\frac{d}{A},\tag{4.3}$$

where $d = 10$ cm is the cell length. Examples of these measurements obtained with a laboratory-blended scintillator sample (using the same composition as the Region I scintillator) are shown in Fig. 4.4a.

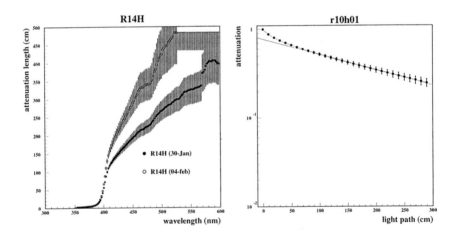

Figure 4.4: Attenuation length vs. wavelength for the Gd-loaded scintillator (of the same type of that used at Chooz) at different aging stages (left) and scintillation light attenuation vs. path (right).

An overall attenuation for the scintillation light was obtained by folding the measured attenuation at different wavelengths with the fluorescence spectra of the phosphores and the PMT photocathode sensitivity; the light attenuation as a function of the path is shown in Fig. 4.4b. Apart from the first ~ 20 cm, needed for the absorption by the second shifter (bis-MSB), the attenuation is well described by an exponential decrease giving an "effective" length. Measurements performed on the laboratory sample at different aging stages reproduced the attenuation length values obtained for the target scintillator within 15%.

4.3.2 Scintillator aging tests

The time variation of the light attenuation length in the detector is well reproduced by the function

$$\lambda(t) = \frac{\lambda_0}{1 + vt} \qquad (4.4)$$

which accounts for the observed exponential decrease of the number of phe's with time; here v is proportional to the velocity of the chemical reactions responsible for the scintillator deterioration. The reaction kinetics is known to depend on temperature according to an exponential law; we then wrote

$$v(T) = v_0 f(T) = v_0 a^{[(T-T_0)/10\,^{\circ}C]} \qquad (4.5)$$

where the index 0 labels the values at room temperature. It is also known that reactions occurring at time scales ~ 1 s are accelerated by a factor ~ 2 if the temperature is raised by $10\,^{\circ}C$ from room temperature.

   We therefore tried to accelerate the aging effects by heating different samples at 60, 70 and $80\,^{\circ}C$. As shown in Fig. 4.5, we found that an increase by 10 degrees in the temperature corresponded to an acceleration by a factor $a \simeq 3$ (instead of 2) in the aging rate. This discrepancy can be explained on the basis of simple thermodynamical arguments if (as it is the case) such a reaction develops at room temperature on a time scale of a few months. The estimated value (at $T_0 = 20\,^{\circ}C$) was $v_0 = (3.8 \pm 1.4) \cdot 10^{-3}\,d^{-1}$, which was acceptable for a 1-year data-taking duration experiment like Chooz. This prediction turned out to be close to the value $((4.2 \pm 0.4) \cdot 10^{-3}\,d^{-1})$ obtained by direct measurements in the detector.

4.4 The photomultipliers tubes

4.4.1 Selection criteria

The PMT's provide both charge and time information needed to determine the event position and the energy deposit. The average photoelectron (from now on phe) yield associated with a typical positron signal ($E \approx 3$ MeV) is ≈ 300, which corresponds to an average ≈ 1.5 phe at each PMT; the number of hit PMT's is $\approx 80\%$ of the total. As PMT's had to work in a single phe regime, the most important feature for tube selection was the single phe resolution, i.e. the peak to valley ratio. Other important requirements were:

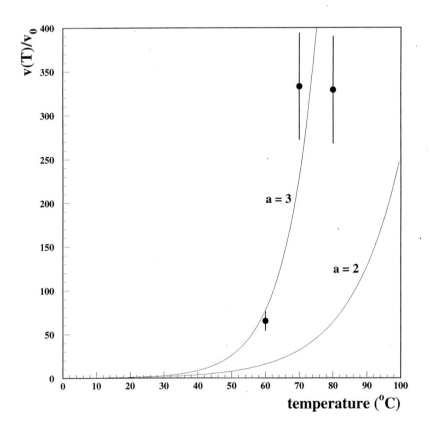

Figure 4.5: Acceleration of the scintillator aging rate as a function of the temperature.

low radioactivity and dark noise, high gain, large photocathode surface, low sensitivity to magnetic fields, good (but not ultra-fast) time properties. The EMI 9536KA B53 [101], an updated version of the 9350 8" diameter 14-dynodes PMT's used in MACRO, turned out to be the best for our needs. The high amplification ($\approx 10^7$, corresponding to $\approx 30\,\text{mV}$ for a 1-phe pulse) matches the characteristics of the front-end electronics; the transit-time jitter ($\approx 8\,\text{ns}$ FWHM at 1-phe level) is comparable with the scintillator decay time ($\approx 7\,\text{ns}$); the PMT bulb is made by an ultra low-radioactivity glass (B53) on explicit request.

PMT's were supplied by the CAEN SY527 HV system (positive anode, grounded cathode), equipped with 16-channel cards A734P and supplying a maximum of 3 kV voltage to the PMT's; the HV channels are CAMAC controlled via the CAENET card C117B. The high voltage is distributed to the dynodes through a "tapered bleeder" progressive divider, with interdynode voltages progressively increasing in the last dynode stages. This divider was preferred to the linear one (with equal interdynode voltages) since it was proved to provide enhanced linearity at a still acceptable high voltage for the required gain. The divider consisted in a circular printed circuit board soldered to PMT socket pins and glued to the back of the socket itself. The divider and the socket pins were enclosed in a Plexiglass cylinder in order to prevent contact of liquid scintillator with all components.

4.4.2   The test facility

A PMT test facility [100] was designed to determine the relevant parameters of all PMT's. The measurement program allowed us to determine the below-listed quantities and to compare the results with the ones contained in the PMT test ticket given by the producer:

(i) the operating voltage corresponding to 30 mV pulse (on 50Ω) for a single phe;

(ii) the PMT noise level;

(iii) the relative light sensitivity (proportional to the photocathode quantum efficiency × phe collection efficiency);

(iv) the single phe pulse height spectrum and its peak to valley (P/V) ratio;

(v) the PMT time characteristics (time jitter and "walk" effects).

The results were used to define the optimal working conditions and decide a proper geometrical arrangement of all PMT's in the detector.

Since the light level of neutrino events in Chooz corresponds to a few phe's at each PMT, it is this regime at which the listed parameters must be determined. The best suited source was the Hamamatsu light pulser PLP-02 with a laser diode head SLD-041 emitting at 410 nm, a wavelength close to the maximum photocathode sensitivity and to the maximum bis-MSB emission wavelength; details about the features of this source, the coupling to the optical bench and to PMT housing structure can be found in [100]. The determination of the PMT relative sensitivity was cross-checked by using two additional "passive" light sources: a disk of NE110 scintillator activated by a low intensity ^{60}Co and a ^{241}Am α-source coupled to a NaI crystal. All our sensitivity measurements were found to be in good agreement with each other and with the quantum efficiency measured by EMI (for details of the measurements see also [102]).

The HV supply system was based on the CAEN SY527 (the same later used in the experiment). The acquisition system used standard CAMAC circuitry (two 16-channel discriminators Phillips 7106, one 32-channel scaler Phillips 7132H, two 16-channel TDC's Phillips 7186H, two 16-channel ADC's CAEN C205A, timing modules, I/O registers, logic units). The data acquisition was driven by LabVIEW codes, running on a Macintosh QUADRA 650, linked with CAMAC libraries as well as with CERN histogramming packages.

4.5 Detector simulation

4.5.1 General description

The Monte Carlo simulation of the detector is based on a Fortran code linked with the CERN GEANT 3.21 package [103]. The program allowed us to simulate the detector response for different particles, in particular for the positron and the neutron following an $\overline{\nu}_e$ interaction. The use of GEANT routines had a two-fold objective:

- define the detector geometry and the physical parameters inherent to the different materials (scintillators, acrylic, steel) the detector is made of:

- track positrons, electrons and photons in the detector and evaluate the energy deposit in the scintillator.

A specially written routine was used to track secondary particles below 10 KeV, the minimum cut-off energy allowed by GEANT, down to 1 KeV, so as to better account for the scintillation saturation effects arising at low energies. The routine calculates the range corresponding to the particle energy and divides this range into 10 steps; the ionization loss is evaluated at each step, and the particle is tracked until its residual energy reaches 1 KeV. Each energy deposit is then converted into the scintillation yield; ionization quenching effects are taken into account according to Birk's law, with $\beta = 1.3 \cdot 10^{-2}$ cm/ MeV derived from the measurements performed on the similar MACRO liquid scintillator [104].

Scintillation photons are isotropically generated and tracked up to the PMT surface. Effects due to light attenuation by the scintillators, due to reflection on the PMT glass, to the light refraction at the target boundary and to the absorption on opaque surfaces (such as the Geode and the calibration pipes) are also simulated. The probability of detecting a phe hitting the PMT surface is weighted according to the quantum efficiency.

The Monte Carlo code (MC) includes also the charge response of digitising electronics; the charge output for each event is arranged in the same format used for real data, which allowed us to reconstruct MC and real events by means of the same algorithm. The evaluation of the neutrino detection efficiencies strongly relies on MC evaluations; it was therefore crucial that the MC output was as close as possible to the data. The most important difference between the real and simulated geometry concerns the shape of the PMT's, assumed to be flat and lying on the geode surface (whereas the PMT glass is hemispherical and protrudes towards the centre by a few cm). As we shall see in Chapter 6, this approximation (needed to avoid a large waste of computing time) is also responsible for the fragility of the reconstruction algorithm in the case of events close to the PMT's.

4.5.2 The neutron transport code

User-defined routines [105] were introduced in the main MC code to follow the neutron moderation and capture processes. A Fortran block data was written with tabulated cross section values for both elastic scattering and capture on Gd, H, C, O, Fe [106]. At each step (whose length depends on the total cross section at the neutron velocity) a decision is taken if either scattering or capture processes (or neither) arise. In the case of captures, γ-rays are emitted in cascade according to the de-excitation scheme of the involved nucleus; each photon is then tracked by GEANT routines. Captures on hydrogen are followed by the emission of one 2.2 MeV γ-ray. The case of gadolinium is harder to manage; the excited nucleus can decay into the ground state through a series of intermediate excited levels, both discrete and continuously distributed. Each capture event typically releases about three γ-rays with a total energy depending on which Gd isotope captures the neutron. The information used about the energy levels involved and the relative decay branches are taken from [107].

Part III

Detector setup and calibration

Chapter 5

Neutrino trigger and data acquisition

The electronic system for the Chooz experiment collects and processes PMT output signals to select events due to neutrino interactions and to separate them from the background. The charge and time information provided by the PMT's are used to reconstruct the energy and position of each event. Let us see how the neutrino trigger operates and how the event information recorded by the digitising electronics is written in the data stream.

5.1 The main trigger

The signature of a neutrino interaction relies on the delayed coincidence of two signals within a $100\,\mu s$ time window: the former, due to the positron, has an average energy of about $3\,\mathrm{MeV}$; the latter, due to the delayed neutron capture on gadolinium, has a total γ-ray energy $\approx 8\,\mathrm{MeV}$. The identification of this event combination is therefore based on a two-level trigger system; the first level selects events with an energy release $> 1\,\mathrm{MeV}$; the occurrence of two first-level triggers, within the $100\,\mu s$ time window, generates the second level which triggers the data digitisation and recording by the on-line system.

5.1.1 The first-level trigger

The electronic sum of the signal corresponding to photoelectrons detected by all PMT's gives a first rough measurement of the energy released by any event. A Monte Carlo simulation of electrons uniformly generated in the detector (whose results are summarized in fig. 5.1) shows that the collected charge is linearly dependent on the energy deposit and almost independent of the position within Region I. On the contrary, the total number of photoelectrons may diverge for events closer than $30\,\mathrm{cm}$ to the PMT's and be even 10 times larger than that obtained at the detector centre. So, a trigger exclusively based on the charge information would not reject lower energy events generated close to the PMT surface.

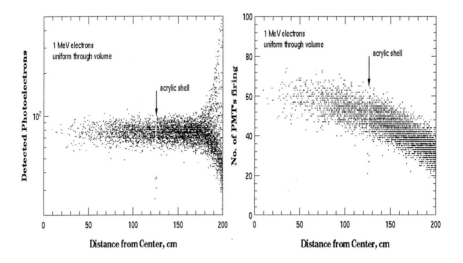

Figure 5.1: Number of detected photoelectrons (left) and number of hit PMT's (right) for 1 MeV electron events, as a function of the distance from the detector centre.

A "topological" cut is then applied by requiring also a minimum number of hit PMT's. Fig. 5.1 clearly shows that this extra condition preferentially rejects events close to the PMT's.

The first-level trigger is generated when both the charge and PMT multiplicity fulfil the above criteria. A scheme of the trigger circuit is presented in fig. 5.2. The signals from each PMT are fed into the front-end electronics made of fan-in/out modules developed for the purpose. These modules provide a linear sum of the input signals (QSUM) whose amplitude is proportional to the total number of photoelectrons. The PMT multiplicity signal (NSUM) is obtained in a similar way; another copy of the PMT signals is fed into the input channels of the Lecroy LRS 4413 discriminators; after setting the threshold to 15 mV ($\approx$ one half of the single photoelectron amplitude), the NSUM signal is obtained from the linear sum of the Σ output of these discriminators, whose amplitude is proportional to the number of channels over threshold on each discriminator board. The first-level trigger is finally asserted when both the resulting QSUM and NSUM signals exceed the preset thresholds.

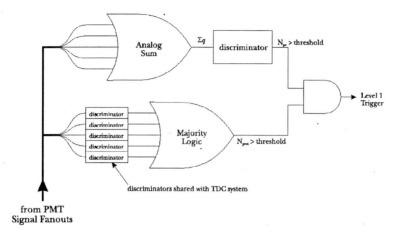

Figure 5.2: First-level trigger scheme. Both the number of photoelectrons and the number of hit PMT's are required to fulfil a certain threshold condition.

5.1.2   The second level trigger

According to what has been stated above, the second-level condition (the one identifying the neutrino interactions) triggers on the occurrence of two events satisfying the first-level condition within a $100\,\mu$s wide time window. This width was chosen to keep the neutrino detection efficiency as high as possible compatible with the accidental background rate. The positron-neutron delay distribution follows an exponential distribution with average $\tau \simeq 30\,\mu$s, so only 5% of the neutrino events are rejected by the delayed coincidence condition.

The second-level trigger (L2) logic is somewhat more complicated than the simple scheme just described. As a matter of fact, there are two different conditions for the fulfilment of the first-level trigger corresponding to two different energy thresholds: QSUM+NSUM low and QSUM+NSUM high. The low condition (L1lo) roughly corresponds to 1.3 MeV and has a rate $\approx 130\,\text{s}^{-1}$, the high condition (L1hi) corresponds to ≈ 3 MeV and has a rate $\approx 30\,\text{s}^{-1}$. The higher threshold discriminates the large background due to the natural γ-radioactivity (which ends with the 2.6 MeV ^{208}Tl line) and is sufficiently low to detect signals due to the neutron capture on gadolinium with almost full efficiency.

The second-level trigger is anticoincided if some activity is present in region III; this rejects events due to the cosmic rays. The VETO condition is on whenever VSUM, the sum of the signals of the 48 PMT's located in the outer region, is over a preset threshold. If no VETO condition occurred during a 1 ms time interval preceeding L2, the L2 triggers: the acquisition

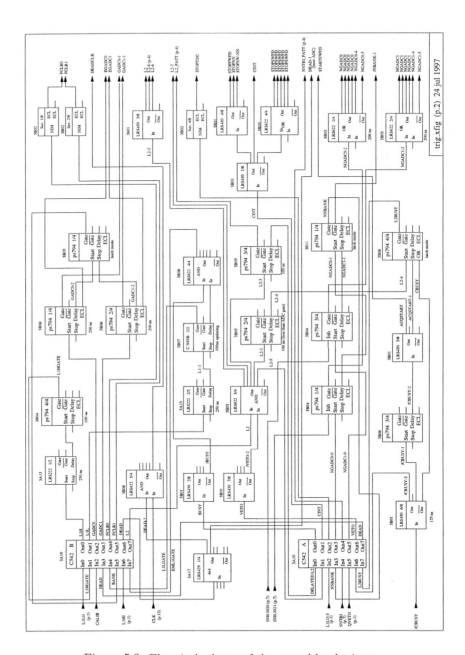

Figure 5.3: Electrical scheme of the second-level trigger.

of further secondary particles goes on for an extra-time of $100 \, \mu s$ and then stops. The on-line processor takes $\approx 80 \, \text{ms}$ to read all the electronics and to record the data on disk. During this time the computer busy condition (*CBUSY*) is on and the acquisition of any further event is disabled. The trigger logic is presented in fig. 5.3.

Although a neutrino interaction corresponds to an L1lo-L1hi or L1hi-L1hi time sequence, the trigger logic enables the acquisition of events with an L1hi-L1lo sequence as well; this allows systematic study of the accidental background.

Finally, the acquisition rate ($\approx 0.15 \, \text{s}^{-1}$), although rather low, is large if compared to the expected neutrino rate at full reactor power (≈ 30 events d^{-1}). The typical event size is roughly 30 Kbytes; the acquisition system must be capable of handling a daily amount of data of the order of 0.5 Gbytes. This capability is achieved through fast readout electronics and large storage devices.

5.1.3 Trigger for calibration runs

Acquisition schemes different from the normal neutrino trigger are required for calibration or particular test runs. The trigger logic can be changed at the beginning of each run by loading an appropriate "trigger table", a combination of the level trigger, veto and busy signals assembled by a dedicated logic unit Caen C542.

Neutron sources

Neutron sources have been extensively used for detector calibration, both to define an absolute energy scale and to measure the neutron detection efficiency. ^{252}Cf is the most frequently used neutron source. This nucleus undergoes spontaneous fission by emitting prompt γ, whose energy is lower than 10 MeV, and prompt neutrons. The neutron multiplicity obeys Poissonian statistics with average value 3.787 [108] and the kinetic energy is Maxwell-distributed, with average 1.3 MeV.

With the source inserted in region I, almost the totality of neutrons are contained and captured: some ($\approx 15\%$) are captured on hydrogen, the majority on gadolinium. The neutron binding energy is released in the form of γ-rays (one 2.2 MeV γ for H, an average of 3 γ's and $\sum_\gamma E_\gamma \approx 8 \, \text{MeV}$ for Gd). In order to record events due to neutron capture on H as well, we defined a different trigger logic bypassing the too strict L1hi threshold condition; the L2 condition relies upon the occurrence of two signals both satisfying the L1lo threshold. The "positron" role is, in this case, played by prompt γ's, while the second L1 trigger is provided by the first neutron capture (if within the usual time coincidence window).

The source is inserted in the detector through two vertical calibration pipes: a central pipe, along the symmetry axis of the detector, and a lateral one, in region II, just at half distance between the geode and the target boundary, as shown in fig. 5.4. It is then possible to study the detector response everywhere along the z-axis in both regions. In particular, as we shall see in the next chapter, calibration runs at different, known, positions allow an accurate tuning of the reconstruction algorithms and comparisons with Monte Carlo method predictions.

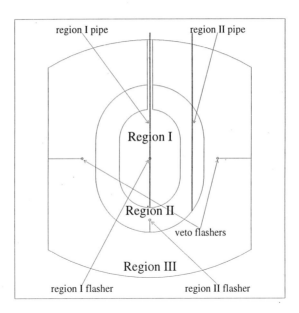

Figure 5.4: Location of laser flashers and calibration pipes in the detector.

γ sources

Daily calibration runs were performed by means of a ^{60}Co source, a well-known γ-emitter providing a low-energy calibration point (2.5 MeV), the sum of 1.17 and 1.33 MeV γ-lines. As we shall see in the following chapters, these runs were particularly useful to follow the detector evolution (a critical point because of the aging of the Gd-loaded scintillator) and tune and check the trigger efficiency.

The neutrino trigger is not suitable to acquire these events, since the delayed neutron signal is missing . In this case the L2 trigger is made to coincide with an L1lo trigger. In such a case, with such a low threshold and without a coincidence request, a great number of events due to γ-radioactivity natural background is recorded.

Laser

A laser based calibration system is also available. It consists of a nitrogen UV Laser emitter ($\lambda = 337$ nm) with a repetition rate ≈ 10 Hz and a time jitter of less than 1 ns between the trigger and the light emission. The light intensity can be selected by using two neutral variable density filter wheels rotated by remote-controlled stepping motors. The attenuated light is fed at six different positions in the detector (as shown in fig. 5.4) by optical fibres. Each fibre end is immersed in a (PPO-based) scintillator kept in a small container (*flasher*), diffusing and shifting the laser light to match the photocathode quantum efficiency. A fraction of the light from each optical fibre is also read by a reference PMT; this way the light intensity is monitored and recorded on a pulse to pulse basis.

In such a case the L2 trigger coincides with the internal laser trigger and is delayed by ≈ 200 ns relative to the light emission.

5.2   Data acquisition

The data acquisition system consists of a set of digitising circuits controlled by a VME-based processor. They are:

NNADC $\rightarrow$ Neural Network Analog to Digital Converter units (VME),
SWFD $\rightarrow$ Slow (20 MHz) WaveForm Digitisers (VME),
FWFD $\rightarrow$ Fast (150 MHz) WaveForm Digitisers (FastBus),
FBADC $\rightarrow$ FastBus Analog to Digital Converter units (FastBus),
FBTDC $\rightarrow$ FastBus Time to Digital Converter units (FastBus).

The PMT signals are fanned out and fed to each digitising equipment, according to the diagram shown in fig. 5.5. For the first three systems, the signals from groups of PMT's are initially fanned in; for the Neural Network ADC's and SWFD's the PMT signals are linearly added by groups of 8, a PMT *patch*. This reduces the number of channels to 24. For the FWFD's the PMT groups varied from 4 to 8 during the experiment.

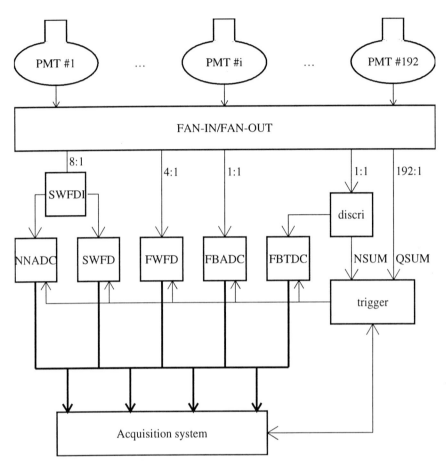

Figure 5.5: Electronics layout for the Chooz experiment, including front-end, trigger and digitising modules.

5.2.1   The on-line system

The acquisition system is based on a heterogeneous apparatus combining different bus standards (see fig. 5.6); the digitisers listed above use both the VME and the FastBus standard, while the trigger electronics (including discriminators and logic units) is based on CAMAC. The information assembling is managed by a VME, OS-9 operating processor, whose central unit consists of a Motorola 68040 microprocessor mounted in a Ces FIC8214 board. After completing the event readout, the data are sent through Ethernet to a dedicated SUN/Unix station and written on disk. A LabVIEW run controller is the interface between the user and the SUN; the controller also provides real-time information about the run and data quality.

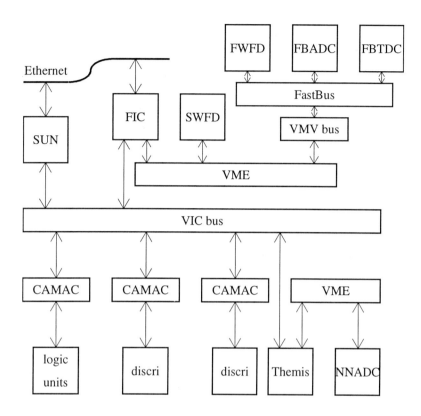

Figure 5.6: On-line system architecture, with special reference to the different bus standards (VME, FastBus, CAMAC) and their interconnection (VIC,VMV).

5.2.2 The digitising electronics

The NNADC's

The Neural Network ADC system is based on a neural network algorithm able to rapidly (200 μs per event) reconstruct the energy and the position of any event in the detector; it also provides a "minimum-bias" trigger for neutrino interactions. Moreover, the digitisation performed by this system is used later for off-line analysis; the first results published by the Chooz collaboration [70] are based on an NNADC data analysis. Let us focus on a description of this digitising electronics, leaving the performance of the NN algorithm and the associated trigger to the next section.

The signals from the 24 PMT patches are sent to two VME, 12-bit ADC banks, each composed of 3 8-channel Caen V465 boards; two banks are necessary to avoid any dead time between positron and neutron pulses, the time needed for the charge integration being $\approx 14\,\mu s$. The charge is integrated within a fast 200 ns gate. An external logic alternatively switches the gate to the banks on the occurrence of an L1lo trigger. The digitised ADC values are arranged in a FIFO, thus allowing an internal multi-hit buffering in each bank. The ADC readout is performed by a Motorola 68030-based Themis TSVME133 VME processor board (referred to as "Themis" in fig. 5.6) which also controls the execution of the neural network algorithm. The TSVME133 is usually in a wait status until it receives a VME interrupt generated as soon as one of the two ADC banks is "not empty".

The NNADC circuitry also includes:

- 1 scaler Caen V560N:

- 1 flash-ADC Caen V534:

- 1 input/output register Caen V513:

- 1 Adaptive Solutions CNAPS parallel processor board.

The scaler and flash-ADC provide redundant pieces of information about the relative timing of events. The scaler is made to count the clock ticks elapsing between each event pair, the clock frequency being 4.84 MHz, so as to obtain the delay of the neutron with respect to the positron. The flash-ADC board (8-bit resolution) is used to sample the trigger signals (L1lo, L1hi, L2 and VETO) at the same clock speed as before. Other channels are devoted to sample the gate pulses to record the ordering of the ADC banks (which is necessary due to the bank switching logic). The page depth amounts to about $850\,\mu s$, from $\approx 750\,\mu s$ before the L2 trigger to $100\,\mu s$ after the L2 trigger.

The last two boards are reserved to the neural network trigger and their function will be later described.

The SWFD's

The Slow WaveForm Digitiser system uses a set of 5 flash-ADC Caen V534 boards, with 20 MHz internal clock speed. These digitisers record the 24 patch plus 6 Veto patch signals from $100\,\mu s$ before to $100\,\mu s$ after the L2 trigger. All input signals need a proper stretching in order to fit to the inherent 50 ns resolution. Therefore these digitisers are essentially peak sensing devices and are not intended for pulse shape analysis. Extra channels record all the detector triggers.

The FWFD's

With their 155 MHz sample clock, these digitisers are the unique equipment to study pulse shapes. These units are equipped with an 8-page memory, the depth of a page being 206 ns, so as to study up to 8 events before the stop.

The FBADC's

The FastBus ADC system has 4 96-channel Lecroy 1885F boards, so as to separately measure the anode charge of each geode PMT. Also in this case, the boards are arranged in two alternating banks in order to record both events associated with the L2 trigger.

The FBTDC's

The FastBus TDC's record the time information of each geode PMT. Referring to fig. 5.5, the input signals consist of logical (NIM) pulses out of the trigger discriminators at minimum threshold. This system consists of 2 96-channel Lecroy 1877 multi-hit units able to record up to 16 hits in a single channel with 2 ns resolution. Their depth has been doubled to 128 μs in order to cover the entire time coincidence window.

5.3 Data structure

The structure of the data coming out of the digitising electronics reflects the characteristics of single equipments; the standard used for data recording must therefore be adapted to the different logic schemes of each data stream, while maintaining the data elaboration time as short as possible.

We are led to introduce a "packet" data structure, where data corresponding to a particular equipment are arranged in a packet, which can be identified by its own code. Other packets, besides the equipment ones, are reserved to store important information concerning acquisition and reactor status. For instance, the run type (neutrino or calibration) is written at the run beginning; packets recording the trigger rates, dead time, reactor power are written at every tenth L2 trigger. The data file of an acquisition run consists therefore of a packet sequence, whose typical order is the following:

begin run packets:	RUN_HEADER
	HV_CONFIG
	DEAD_CHANNELS
event # 1:	EVT_BEGIN
	SWFD
	NNADC
	FWFD
	FBADC
	(continued on next page)

(continued from previous page)
FBTDC
EVT_END

⋮

event # 10:   EVT_BEGIN
SWFD
NNADC
FWFD
FBADC
FBTDC
SCALER
REACTOR_INFO
EVT_END

⋮

end run packets: RUN_END
RUN_SUM

5.3.1 The "NNADC" packet

5.3.2 The Reactor Power data

As explained in Chapter 3, we have two current loops at our disposal (one per reactor) whose amplitude is proportional to the neutron detection rate inside the reactor core. The voltage measured across a 5Ω resistor (connected in series to each loop) provides direct information about the thermal power for both reactors. Such a measurement is made by a CAMAC, 9-bit resolution voltmeter. The voltmeter calibration is shown in fig. 5.7, where thermal power values, provided by E.d.F., are plotted versus the corresponding voltmeter data; the calibration parameters obtained by a linear fit are shown also.

5.4 The neural-network based trigger

5.4.1 Motivations

The use of a Neural Network (NN) based trigger had a twofold objective. First, it was intended as a fast acquisition filter for selecting events inside a

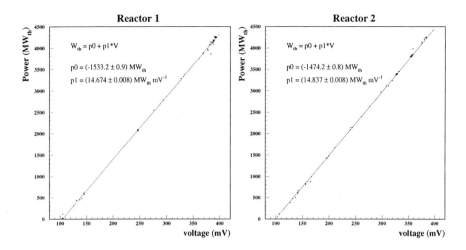

Figure 5.7: Voltmeter calibration for reactor 1 (left) and 2 (right).

fiducial volume, thus reducing the background and the acquisition rate. The major background component is due, as expected, to accidental coincidences of γ radioactivity events, mostly concentrated near to the PMT surface. A large fraction of these events can then be rejected by requiring a spatial correlation between the positron and the neutron.

Secondarily, it could be used for loosening the L2 trigger condition. As in the case of neutron source calibrations, the idea is to require a second-level trigger upon coincidence of two L1lo triggers, so as to recuperate neutrino events accompanied by a neutron capture on hydrogen. Such a technique is not essential in the case of Chooz, since the amount of H-captures is $\approx 15\%$ of the total neutrino interactions in the target. On the contrary, it might play a crucial role in future reactor experiments (namely Kamland), where the plan is to use only a high-flash liquid scintillator (not Gd-loaded). Also in Kamland the radioactivity is expected to be concentrated in the outer shell of the detector, so an on-line cut, based on the relative positron-neutron distance, would be much more effective in a detector whose linear dimensions are one order of magnitude larger than Chooz.

5.4.2 Hardware implementation

The NN algorithm is executed by the CNAPS/VME board, equipped with a 64-node parallel processor [109]. These nodes are arranged to run a standard back-propagation algorithm, with 60 nodes at the hidden layer and 4 (as many as the event coordinates) at the output layer. The 24 input ADC values are loaded into the CNAPS RAM memory by the Themis processor, which also controls the execution of the algorithm. The algorithm works with an 8-bit resolution for input/output data and with 16-bit weights.

The code for running the CNAPS under OS-9 was obtained by modifying the factory supplied standard software (running under VXWORKS). Several further modifications were introduced to fasten the event reconstruction. The delay between the gating of one ADC bank and the complete event reconstruction was measured to be $\approx 180\,\mu s$. Most of this time ($\approx 100\,\mu s$) is spent for the VME interrupt handling. ADC readout and data transfer to the CNAPS memory take an additional $70\,\mu s$, so that the time needed for the algorithm execution ($\approx 10\,\mu s$) is almost negligible in comparison.

After the event reconstruction, the NN trigger searches for pairs of events to select neutrino interactions followed by neutron capture on hydrogen. A good efficiency for these events is achieved by the following requirements:

- distance from the PMT edge larger than 20 cm (for both positron and neutron):

- relative distance < 150 cm:

- relative delay < $360\,\mu s$:

As soon as these conditions are fulfilled, an output register on the Caen V513 board is set true and the information is sent to the main trigger. If the L2 trigger is not present and no activity is found in the Veto, the NN trigger is set at on and the acquisition is stopped $100\,\mu s$ later for the NNADC readout.

The NN average rate was measured to be $\approx 0.3\,s^{-1}$ (about twice larger than the L2 rate). The event size is small (1 Kbytes) so that the introduction of the NN trigger increased the data yield by about 7% and the dead time fraction from 4% to 5.7%.

5.4.3   Network training

The network has been trained for 100 epochs on 10000 Monte Carlo electron events generated according to a flat energy distribution up to 10 MeV and uniformly distributed in the geode volume. The results of the training (which takes about 15 min.) are shown in fig. 5.8 where the difference between the network reconstructed and the Monte Carlo generated energy and position is plotted. The Gaussian fit superimposed on the z distribution gives a resolution $\approx 15\,cm$ (the same for the other two coordinates) to be compared with the 6 cm obtained by means of the standard minimization procedure used in the offline analysis (which we shall deal with in the next Chapter). The tail of underestimated energies is due to events generated near the acrylic

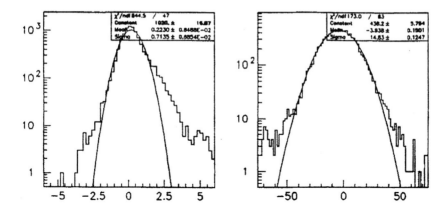

Figure 5.8: Monte Carlo generated minus CNAPS reconstructed energy (left) and z-position (right) for electron events.

vessel, since, in this case, part of the energy is lost in the acrylic. The energy resolution turns out to be 0.71 MeV, while the standard fitting procedure obtains 0.46 MeV.

The complete system described above was hardware tested by means of a CAMAC DAC system in conjunction with a linear gate generator. The CAMAC was driven by LabVIEW on a Macintosh via a Bergoz interface. This system could reproduce Monte Carlo generated events at the ADC inputs, thus providing a tool for testing the complete reconstruction chain. We checked that the resulting energy and position were only slightly worse than those obtained by feeding the CNAPS directly with the Monte Carlo events.

Chapter 6

Detector calibration and event reconstruction

The reconstruction methods are based, for each event, on digitisation, performed by the VME ADC's, of the signals from 24 patches; these patches group the 192 PMT's looking at the inner detector region. It was checked that this grouping does not significantly affect the quality of the reconstruction.

A good determination of the position and energy of each event and of the detection efficiency strongly depends on the knowledge of the main parameters of the detector (scintillator light yield and attenuation length, PMT and electronics gains). So, before dealing with the event reconstruction procedures, we must discuss the methods to determine these parameters and their time evolution.

6.1 PMT gain

The stability of the PMT gain was checked by periodic measurements of the single photoelectron pulse height spectrum of each PMT. Each spectrum was obtained by recording the PMT counting rate, out of a discriminator, as a function of its threshold. The electronics for this measurement is presented in fig. 6.1. The data were taken by a LabVIEW code running on the Chosun1 station, in parallel with the main acquisition. No artificial light source was needed, since the single counting rate, at the one-phe level, due to the γ-radioactivity and to the dark noise, was $\approx 1\,\mathrm{kHz}$. Examples of these curves are shown in fig. 6.2. The single phe spectra, obtained by differentiating the counting rate curves, were then fitted to get the PMT gains; values around 30 mV for the geode PMT's and 5 mV for the Veto ones were needed to match our front-end electronics. Linear amplifiers (see Fig.6.1) were used to measure the Veto PMT gain, whose single phe pulse height was lower than

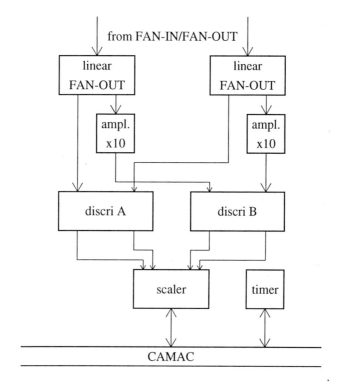

Figure 6.1: Scheme of the electronic chain used to measure the pulse height spectrum of sixteen PMT's at a time; each line feeds the signal from 8 PMT's. The use of amplified channels is needed for setting the gain of the Veto PMT's.

the minimum discriminator threshold. Fig. 6.3 shows the distribution of the average single phe pulse height for the whole of the geode PMT's and its evolution since the beginning of the data taking. One concludes that the PMT's stabilized about one month after burn-in and the asymptotic gain was ≈ 10% lower than the starting value.

6.2   Determination of the photoelectron yield

A similar method was used to determine the photoelectron yield for events at the detector centre. We used the laser flasher at the detector centre as a calibrated light source and an acquisition chain very similar to that shown in fig. 6.1. In such a case, however, the laser itself provides its own trigger pulse about 200 ns later than the light emission, so that the scalers may be gated in coincidence with each laser shot. The contribution to the counting rate due to radioactivity is then negligible.

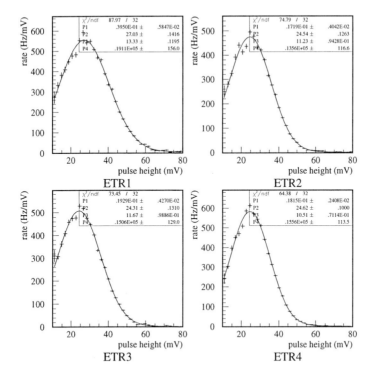

Figure 6.2: Pulse height spectra for a sample of four PMT's.

An estimate of the number of photoelectrons N^i detected by each PMT comes from the probability of having zero phe's when the average number is N^i. From Poisson statistics, we obtained for the i-th PMT

$$N^i = -\log(1 - \frac{N^i_{hit}}{\varepsilon^i N_{shot}})$$

(6.1)

where N^i_{hit}, N_{shot} are respectively the counts over threshold for that PMT and the number of laser shots. (6.1) includes also a detection efficiency ε^i, equal to the probability of single phe pulse height over threshold, which can be evaluated for each PMT from the pulse height spectra shown in fig. 6.2. The number of phe given by this method at different thresholds is presented in fig. 6.4. The light intensity was kept at a level of about 0.5 phe/PMT so as to increase the precision of the method. This number was divided by the energy deposit corresponding to this light intensity in order to determine the absolute light yield. As an energy reference we used the 2.5 MeV ^{60}Co "sum" line. By averaging the number of photoelectrons all over the PMT's, we obtained (0.65 ± 0.03) phe/MeV/PMT and a yield of (125 ± 5) phe/MeV.

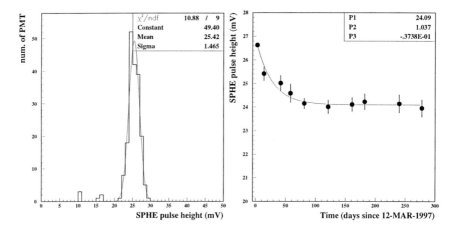

Figure 6.3: Distribution of the single phe peak for all PMT's (left) and its time evolution since the start of data taking (right).

6.3   ADC calibration in a single photoelectron regime

Calibration runs utilizing the laser at a low intensity were periodically performed to test the single photoelectron gain and to calibrate the VME ADC's. The good ADC resolution and the high PMT gain allow us to distinguish the contribution to the ADC spectra due to different number of phe's. A sketch of these spectra is presented in fig. 6.5. The fit function results from the sum of a pedestal and a number of phe-distributions; each of these terms is represented by a Gaussian function (weighted according to Poisson statistics) whose peak position is linear with the number of photoelectrons.

Apart from the ADC calibration, the fit provides also an independent determination of the number of phe's collected by each PMT patch which is in good agreement with the measured photoelectron yield.

6.4   Light attenuation in the Gd-loaded scintillator

A measurement of the light attenuation in the Gd-loaded scintillator was periodically performed throughout the acquisition period. The need of such a test comes from the observation of a slight decrease (roughly exponential) of the charge associated with a calibration source line, as seen in fig. 6.6. As

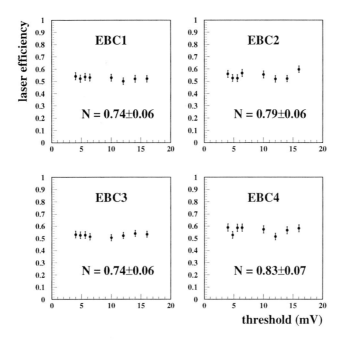

Figure 6.4: Laser efficiency and average number of photoelectrons as a function of threshold for a four PMT sample.

neither the light yield nor the Gd loading showed any observable degradation over time, we were forced to follow the time evolution of the scintillator transparency which is subject to aging.

The method consisted in displacing a light source along the calibration pipe and recording the charge detected by the top and bottom PMT patches. If we assume an exponential light attenuation (which is a good approximation for a light path longer than 10 cm, as remarked in chapter 3), these charge values are related by the simple expression

$$\frac{Q_T}{Q_B} = \frac{\Omega_T}{\Omega_B} \exp(\frac{d_T - d_B}{\lambda_{Gd}}) \tag{6.2}$$

$\Omega_{T,B}$ being the solid angle subtended by the top (bottom) patch PMT's and $d_{T,B}$ the average distance between these PMT's and the source position. The proper light source for this job is the laser; unfortunately, the angular distribution of the laser light is not isotropic. The fibre end is coupled to the glass bulb of the flasher by a 1.5 cm diameter teflon connector. The flasher is oriented with the teflon connector facing towards the top side, therefore

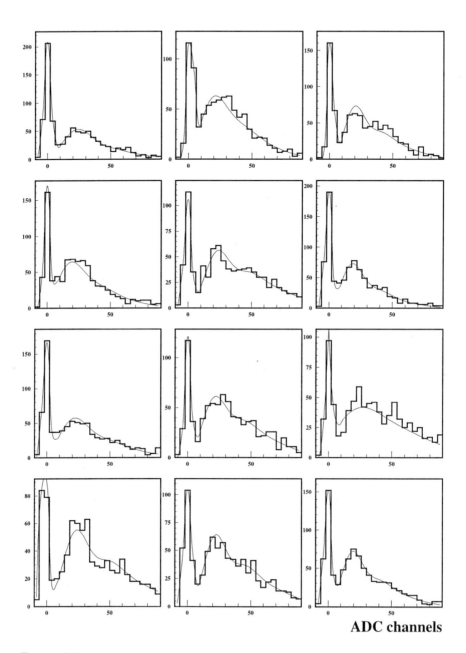

ADC channels

Figure 6.5: ADC spectra obtained with a laser calibration, where the single phe peak is clearly visible. The fitting function results from the sum of the pedestal and the first three phe's.

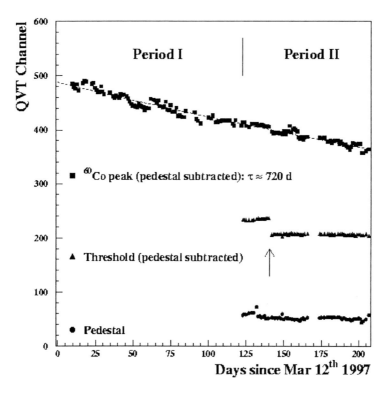

Figure 6.6: Peak associated with the ^{60}Co 2.5 MeV line, as a function of time, as measured by means of a Lecroy QVT. The detected charge follows an exponential decrease, with decay time ≈ 720 d.

generating a shadow effect in the upward direction (this effect is also visible in reconstructing laser generated events, as discussed at §6.6.2).

We were thus forced to use a radioactive source, namely ^{252}Cf. This source was preferred to a γ source (such as ^{60}Co) since the double signature, provided by the prompt γ's and the neutron captures, makes the identification of source events much easier. As a drawback, we had to give up the simple point-like source approximation and evaluate the solid angles and the distances involved by means of a Monte Carlo simulation of the ^{252}Cf neutron emission and capture.

The results obtained by different measurements are displayed in fig. 6.7, superimposed on the exponential best-fit curves according to (6.2). The charge values in use are the ones corresponding to the 2.2 MeV γ-line due to the neutron capture on hydrogen. We did not use the 8 MeV line in order to reduce the systematic effects due to ADC saturation (which might arise when the source approaches the edge of region I). The fitted λ_{Gd} values are

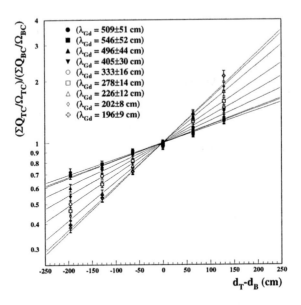

Figure 6.7: Attenuation length measurements at different stages of the data taking period.

plotted versus time in fig. 6.8. The time evolution was fitted by the empirical function

$$\lambda_{Gd}(t) = \frac{\lambda_0}{1 + \alpha t} \qquad (6.3)$$

which accounts for the observed exponential decrease of the number of phe's with time (see fig. 6.6). By taking the λ_{Gd} values, at the beginning and at the end of the experiment, a 35% reduction in the photoelectron yield is estimated (which is that expected from the time behaviour of the QSUM) for events at the detector centre.

6.5 Electronics amplification balancing

The electronic gain may differ from patch to patch and slightly vary with time because of the time evolution of the behaviour of the active electronic components (fan-in, fan-out) following the PMT's. Amplification balancing factors were obtained three times a week by using the ADC charge values corresponding to the 8 MeV line peak generated with the ^{252}Cf source at the detector centre. The dependence of these charge values on the scintillator

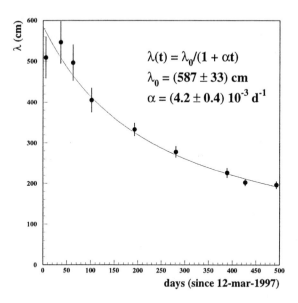

Figure 6.8: λ_{Gd} versus time with best-fit function superimposed.

attenuation and on the detector geometry (solid angles and distances between the source and the PMT's) was unfolded according to the method explained in the previous section.

6.6 Event reconstruction techniques

6.6.1 The standard minimization algorithm

The standard algorithm uses a maximum likelihood method to reconstruct the energy E and the vertex $\vec{x}$ of an event. The likelihood is defined as the joint Poissonian probability of observing a measured distribution of photo-electrons over the 24 patches for given $(E, \vec{x})$ coordinates in the detector. So, for an event occurring at time t after the start of data taking, we can build a likelihood function as follows:

$$\mathcal{L}(N; \overline{N}) = \prod_{j=1}^{24} P(N_j; \overline{N}_j(E, \vec{x}, t)) = \prod_{j=1}^{24} \frac{\overline{N}_j^{N_j}}{N_j!} e^{-\overline{N}_j} \qquad (6.4)$$

where N_j is the observed number of photoelectrons and $\overline{N}_j$ the expected one for the j-th patch given an event $(E, \vec{x}, t)$. The reason for using a Poissonian instead of Gaussian statistics is due to the frequent occurrence of a low energy events with low number of photoelectrons detected by some PMT patches.

The values N_j are obtained from the data recorded by the VME ADC's by applying the reference ADC gain g_0 and the balancing factors f_j: so

$$N_j = \frac{Q_j}{g_0 f_j(t)} \tag{6.5}$$

The predicted number of photoelectrons for each patch is computed by considering a local deposit of energy, resulting in a number of visible photons which are tracked to each PMT through the different attenuating Region 1 ar.d 2 scintillators. Therefore

$$\overline{N}_j = \alpha E \eta \sum_{k=1}^{8} \frac{\Omega_{jk}(\vec{x})}{4\pi} \exp\left(-\frac{d_1^{jk}(\vec{x})}{\lambda_{Gd}(t)} - \frac{d_2^{jk}(\vec{x})}{\lambda_{Hi}}\right) \tag{6.6}$$

where

E is the ionization energy deposited in the scintillators,
α is the light yield of the scintillator,
η is the average PMT quantum efficiency,
Ω_{jk} is the solid angle subtended by the k-th PMT from the event position,
d_1^{jk} is the path length in region I,
d_2^{jk} is the path length in region II,
λ_{Gd} is the attenuation length in region I scintillator,
λ_{Hi} is the attenuation length in region II scintillator.

For computing speed reasons PMT's are considered to be flat and the solid angle is approximated by the following expression

$$\Omega_{jk} = 2\pi \left(1 - \frac{d_{jk}}{\sqrt{d_{jk}^2 + r_{PMT}^2 \cos\theta}}\right) \tag{6.7}$$

r_{PMT} being the PMT photocathode radius and θ the angle between the event-PMT direction and the inward unit vector normal to the PMT surface.

Instead of (6.4), as is usually the case for problems involving the maximum likelihood method, it is more convenient to use the theorem on the "likelihood ratio test" for goodness-of-fit to convert the likelihood function into the form of a general χ^2 statistic[110]. We let N_j be the best estimate of the true (unknown) photoelectron distribution and form the likelihood ratio λ defined by

$$\lambda = \frac{\mathcal{L}(N; \overline{N})}{\mathcal{L}(N; N)} \tag{6.8}$$

The "likelihood ratio test" theorem states that the "Poissonian" χ^2, defined by

$$\chi^2 = -2\log\lambda = 2\sum_{j=1}^{24}[\overline{N}_j - N_j + N_j\log(\frac{N_j}{\overline{N}_j})], \tag{6.9}$$

asymptotically obeys a chi-square distribution[111]. It is easy to prove that the minimization of χ^2 is equivalent to maximization of the likelihood function, so that the χ^2 statistic may be useful both for estimating the event characteristics and for goodness-of-fit testing.

We used the MIGRAD minimizer provided by the CERN-MINUIT package [112] to minimize (6.9). The search for the minimum χ^2 proceeds through the computation of the first derivatives of (6.9). This routine revealed itself as very powerful, provided that the starting values for the fit parameter are accurate. We studied the χ^2 profile by reconstructing Monte Carlo generated events. Several relative minima were found, most of them differing by more than 1σ from the generated $(E, \vec{x})$ coordinates. This is the reason why a suitable choice of these starting values is crucial in event reconstruction. In our case, an indication of the event position comes from the charge distribution of the patches; for instance, no significant asymmetry will be visible for events at the centre, whilst the charge distribution will be more and more asymmetric for events approaching the detector boundary. We therefore grouped the PMT patches into 6 "superpatches" according to detector frame axes as follows:

x-axis	west	$\rightarrow$	east
	WTL		ETL
	WTR		ETR
	WBL		EBL
	WBR		EBR
y-axis	south	$\rightarrow$	north
	STL		NTL
	STR		NTR
	SBL		NBL
	SBR		NBR
z-axis	bottom	$\rightarrow$	top
	NBC		NTC
	WBC		WTC
	SBC		STC
	EBC		ETC

The starting point for the i-th coordinate was finally defined according to the following:

$$x_{i0} = \frac{\sqrt{Q_+^i} - \sqrt{Q_-^i}}{\sqrt{Q_+^i} + \sqrt{Q_-^i}} D^i, \quad i = 1, 2, 3, \tag{6.10}$$

where the indices $+-$ refer to the opposite superpatches of the i-th axis and D^i is the half size of the detector along that axis. Once the x_{i0} corresponding to the starting position is known, the starting energy value is obtained from (6.6) after replacing $\vec{x}$ with $\vec{x}_0$ and $\overline{N}_j$ with N_j.

6.6.2 Performance

The described method was tested by analysing events generated with calibration sources in various positions inside the detector. Let us review the results.

Laser

Distributions of reconstructed events generated by the laser flasher at the detector centre are presented in fig. 6.9. The standard deviation of the distributions shown gives an indication of the resolution, both in energy and position. The fit yields $\sigma_x \approx 4\,\mathrm{cm}$ for each coordinate and an energy resolution $\sigma_E \simeq 0.33\,\mathrm{MeV}$ at an equivalent energy of 8.1 MeV, which is dominated by the statistical fluctuations of the number of photoelectrons. This number also affects the position resolution; the effect is clearly visible in the data taken with the same laser flasher for three different light intensities (in fig. 6.9 a larger filter attenuation corresponds to a lower intensity). As one can see, the average z coordinate is displaced by 7 cm from the nominal position; this is due to the effect of the shadow of the flasher we already mentioned; in this case, the flasher points downward, thus getting a shadow in the upward direction and displacing the event below the true source position.

^{252}Cf source

In figs. 6.10, 6.11, 6.12 we report the results of the calibration runs with the ^{252}Cf source at different positions ($z = 0, -40, -80\,\mathrm{cm}$) along the calibration pipe. The position distribution is everywhere Gaussian for all the coordinates, with $\sigma_x \approx 19\,\mathrm{cm}$. An equivalent number of neutrons was generated, at each position, by our Monte Carlo code and similarly analysed. The figures also display the Monte Carlo data to emphasize the agreement between data and expectations.

The energy spectra show the energy lines due to neutron capture on Gd and H. The shape of the Gd-capture energy spectrum, as pointed out in

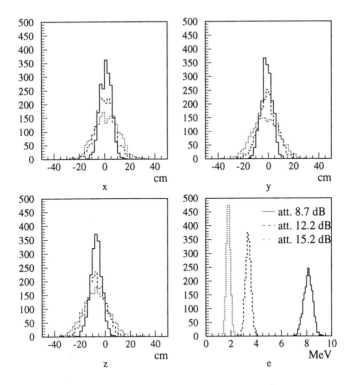

Figure 6.9: Comparison of position and energy distributions for runs with the laser flasher at the detector centre, corresponding to three different light intensities.

fig. 6.13, results from the superimposition of γ lines due to neutron capture on ^{157}Gd and ^{155}Gd (see tab. 4.1 for a reference). The fitted peak values are $\approx 2.5\%$ than the nominal ones, as an effect of the scintillator saturation.

6.6.3 Problems in reconstruction

We have just seen that event reconstruction gives good results, for calibration runs, as long as the source position is inside Region I; it is again worth stressing a very satisfactory agreement with Monte Carlo predictions. This is not unfortunately true everywhere in the detector, as more and more problems arise for events closer and closer to the geode surface.

The main problem is related to the $1/r^2$ divergence of the light collected by one PMT; an exponential light attenuation, entering formulae (6.2,6.6), becomes inaccurate as an event approaches the PMT's. In such a case, also the approximation of a flat PMT surface is no longer adequate to evaluate the solid angle by means of (6.7). These effects are shown in fig. 6.15. The neutron capture energy (both for gadolinium and hydrogen) is overestimated

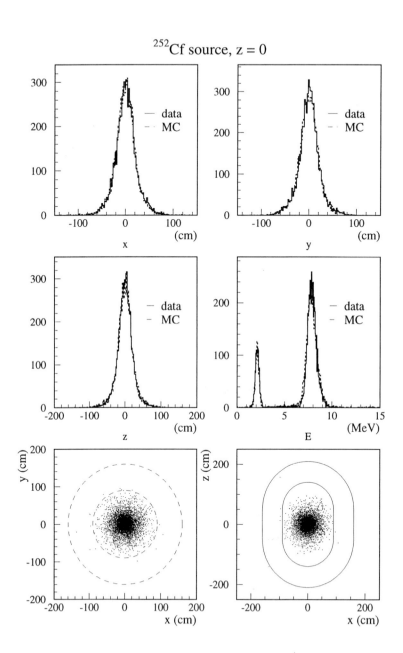

Figure 6.10: Data and Monte Carlo distributions of neutron events with the ^{252}Cf source at the detector centre.

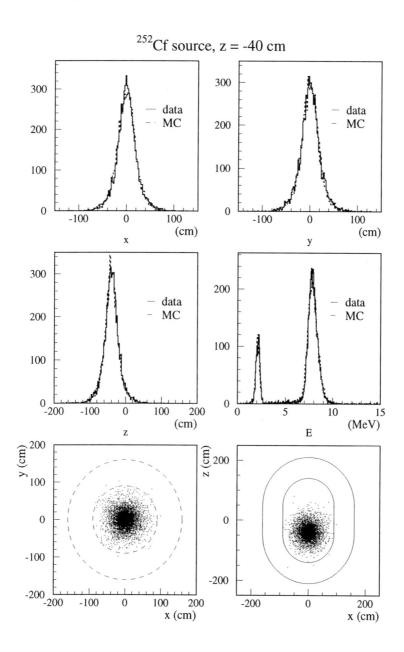

Figure 6.11: Same as before, with the source at $z = -40$ cm.

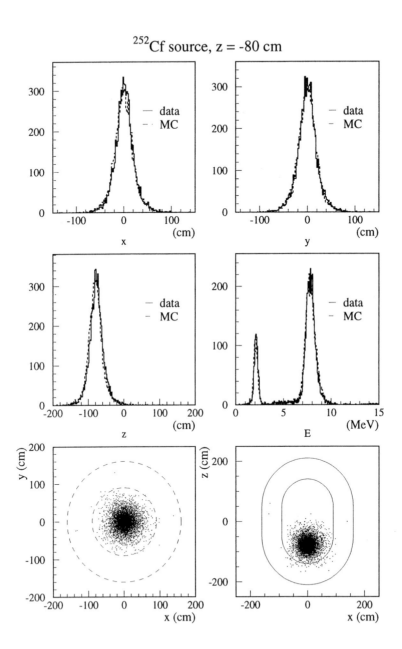

Figure 6.12: Same as before, with the source at $z = -80\,$cm.

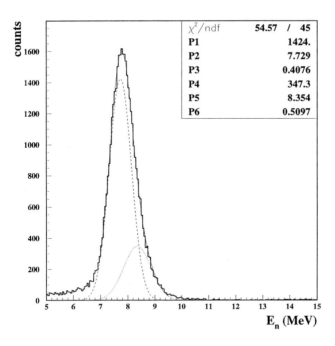

χ^2/ndf	54.57 / 45
P1	1424.
P2	7.729
P3	0.4076
P4	347.3
P5	8.354
P6	0.5097

Figure 6.13: Reconstructed energy spectrum for events associated with neutron capture on Gd. Contributions from γ-lines at 7.94 MeV (capture on ^{157}Gd) and 8.54 MeV (capture on ^{155}Gd) are singled out. The double-Gaussian fit parameters are also shown.

at reconstructed distances smaller than 30 cm from the geode surface.

Fig. 6.15 also shows one further weakness inherent in the minimization procedure. As remarked above, the MIGRAD minimizer heavily depends on the knowledge of the first derivatives of (6.9) with respect to the fit parameters and fails if this computation is not accurate enough. This situation arises for events near the acrylic vessel, where the first derivatives of (6.9) exhibit a discontinuity due to the different light attenuation of the Region I and Region II scintillators; as a result, a discontinuity in the energy vs. distance profile is found around the vessel surface position. A slight improvement was obtained by calling the CERN SIMPLEX routine when MIGRAD failed. This minimizer, although much slower and in general less reliable than MIGRAD, is more suitable in this case since it does not use first derivatives and it is not so sensitive as MIGRAD to gross fluctuations in the function value.

Fig. 6.16 shows the distribution of ^{252}Cf neutron events generated at $z = -120$ cm. The discontinuity in the z distribution is also present for

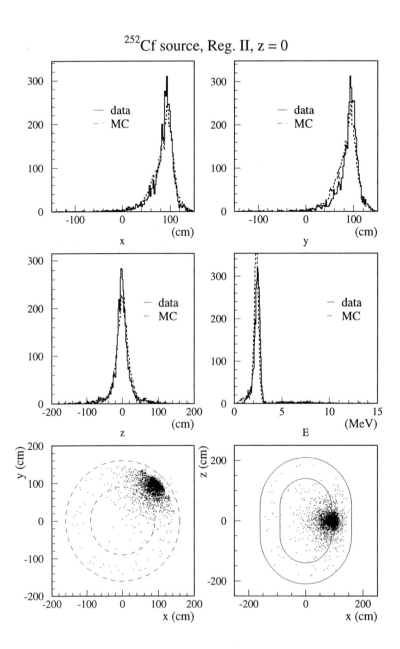

Figure 6.14: Reconstruction of neutron events with the ^{252}Cf source in region II, $z = 0$.

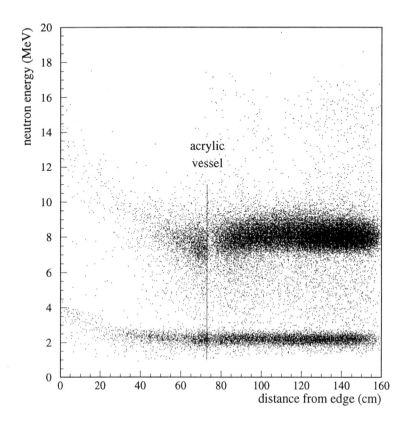

Figure 6.15: Energy versus distance from edge for neutron events generated along the calibration pipe. The energy is flat until the reconstructed position is more than 30 cm far from PMT's.

Monte Carlo generated events, thus proving that this is an inherent feature of the minimization procedure.

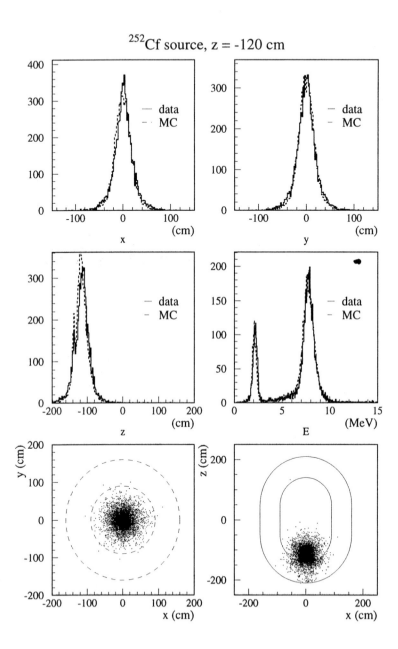

Figure 6.16: Distributions of neutron events with the ^{252}Cf source at $z = -120$ cm. The discontinuity in the z distribution at the vessel surface is visible also in Monte Carlo generated events.

Part IV

Data analysis

Chapter 7

Neutrino signal and background

Events due to neutrino interaction must be selected in a much wider data sample, essentially made of background events. Off-line selections, mostly based on the event reconstructed variables, were then chosen to enhance the neutrino signal and to reduce the background; the final ratio of the neutrino rate at full reactor power vs. the residual background is > 20. A proper interpretation of these results, in terms of neutrino oscillations, requires a careful evaluation of the efficiencies associated with the selection criteria. The analysis procedure is reviewed and the distributions of the final candidate sample are presented. Finally, we will perform an important quality check on the data, i.e. the location of the neutrino source by means of the reaction (3.13); we will also discuss the possible extension of such a technique to Supernova detection.

7.1   The data sample

The data acquisition period extended over 450 days, from March 12th 1997 till July 20th 1998; about 2000 runs were taken (the run number ranging from 579 to 2567), including standard neutrino runs (about 1/4 of the total) and daily calibrations. The experiment stopped taking data five months after the last reactor shut-down (occurred on February 8th 1998) when it became clear that, due to problems related with the cooling system, neither reactor would resume its normal operating conditions for at least one year.

The data taking is summarized in Tab. 7.1; the total thermal energy released (which can be intended as a neutrino integrated luminosity) is also listed. The total live time amounts to $\approx 340\,\mathrm{d}$, 40% of which spent with both reactors off.

Table 7.1: Summary of the Chooz data acquisition cycle from April 7th 1997 till July 20th 98.

	Time (h)	$\int W \, dt$ (GWh)
Run	8761.7	
Live	8209.3	
Dead	552.4	
Reactor 1 only ON	2058.0	8295
Reactor 2 only ON	1187.8	4136
Reactors 1 & 2 ON	1543.1	8841
Reactors 1 & 2 OFF	3420.4	

7.1.1   Power evolution

The power evolution of the Chooz reactors is illustrated in Fig.7.1. The set of power values for both reactors almost continuously covers the entire range up to full power. It is then possible to correlate the neutrino rate to the running thermal power so as to simultaneously determine the net neutrino yield and the background rate. The procedure will be discussed later on. It is now worth stressing that this is a very unique feature of Chooz reactors.

It must be also noticed that Chooz reactors were alternatively off for at least 80% of the total live time of the experiment. This allowed us to extract the contribution to the neutrino signal from separate reactors and to perform, thanks to their different distances from the detector, a two-distance oscillation test.

7.2   Neutrino selection

7.2.1   Preliminary selection and event classification

A primary reduction of the radioactivity background events is obtained by applying loose energy cuts to the neutron-like signals; events are selected if $QSUM_n > 13000$ ADC counts, which roughly corresponds to a 4 MeV energy deposit at the detector centre. The residual events (≈ 720000 over a total number of ≈ 12000000 L2 triggers) are then reconstructed by the standard minimization procedure described in the previous Chapter.

An analysis of this preliminary sample gives a first illustration of the properties of the neutrino signal (reactor-on data) and the associated background (reactor-off data). Figs. 7.2,7.3 show the correlation of neutron-like vs. positron-like energy for this sample. Neutrino events, as indicated in Fig. 7.2, populate a region in the (E_{e^+}, E_n) plot delimited by $E_{e^+} < 8$ MeV

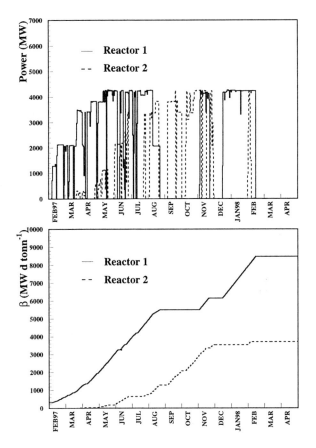

Figure 7.1: Power (top) and burn-up (bottom) evolution for Chooz reactors. Both have been off since February 1998.

and $6 < E_n < 12\,\mathrm{MeV}$. Background events, depending on their position in these scatter plots, are classified in the following categories:

A) events with $E_{e^+} < 8\,\mathrm{MeV}$ and $E_n > 12\,\mathrm{MeV}$:

B) events with $E_{e^+} > 8\,\mathrm{MeV}$ and $E_n > 12\,\mathrm{MeV}$:

C) events with $E_{e^+} > 8\,\mathrm{MeV}$ and $6 < E_n < 12\,\mathrm{MeV}$:

D) events with $E_{e^+} < 8\,\mathrm{MeV}$ and $E_n < 6\,\mathrm{MeV}$:

The neutron energy distribution of C) events presents a clear $8\,\mathrm{MeV}$ peak, typical of the neutron capture on gadolinium, still persisting in the reactor-off data. These events can then be interpreted as a correlated background associated with high energy spallation neutrons originated from cosmic ray

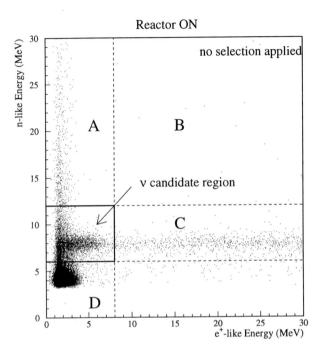

Figure 7.2: Neutron versus positron energy for neutrino-like events collected during the reactor-on period. A preliminary cut to the neutron QSUM is applied to reject most of the radioactivity background.

interactions in the rock surrounding the detector; the neutrons entering the detector are slowed down to thermal velocities via elastic scattering on protons and then captured; the proton recoil signal, whose energy spectrum is roughly flat and extends over 50 MeV, mimics the positron signal. This interpretation is confirmed by the neutron delay distribution; an example is shown in Fig. 7.4; events in sample C) follow an exponential decay distribution whose life time $\tau = (30.5 \pm 1.0)\,\mu s$ is peculiar of the neutron capture in the Gd-doped scintillator.

Events in categories A) and D) show a flat delay distribution, therefore implying an accidental coincidence of uncorrelated signals. In both cases, the positron signal is faked by a low energy $(E_\gamma \leq 3\,\mathrm{MeV})$ radioactivity event. The neutron signal may be associated with either a radioactivity event, as in case D), or a high activity signal (most likely due to proton recoil) as in case A).

Finally, the delay of B) events is exponentially distributed; the lifetime $\tau = (2.8 \pm 0.4)\,\mu s$ is compatible with the lifetime of muon decays at rest. So,

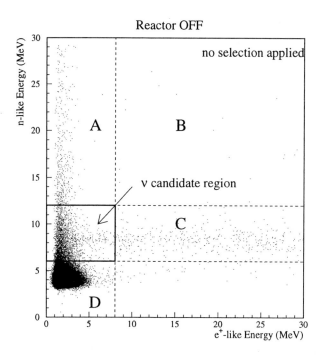

Figure 7.3: Same as before, with neutrino-like events collected during the shutdown of both reactors.

they can be associated with residual cosmic muons stopping in the detector and then decaying. In such a case, both the muon energy loss and the Michel electron energy are much higher than what is typical in a reactor neutrino interaction; these events can be rejected by applying an energy selection both to the positron-like and the neutron-like signals.

The accidental background can be significantly reduced by applying fiducial volume cuts. Less than 10% of the events in regions A) and D), as seen in Fig. 7.5, survive the selection cuts (distances from the geode $d > 30$ cm for both positron and neutron and relative distance < 100 cm). The correlated background instead is more difficult to eliminate as these events exhibit the same features of neutrino interactions. This is the reason why the final background rate is dominated by the correlated component, in spite of the shielding provided by the rock overburden. However, these neutron signals are an important tool to follow the energy calibration stability throughout the experiment. The results are shown in Fig. 7.6, where the average neutron energy, obtained by a Gaussian fit of the gadolinium capture peak, is plotted vs. the run number. The stability throughout the data taking period

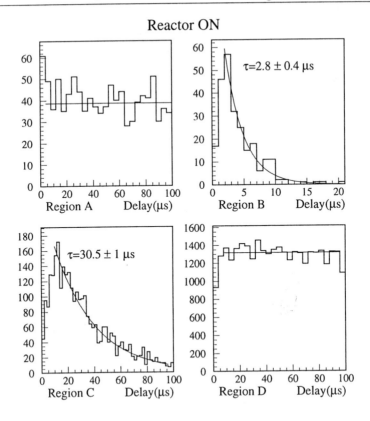

Figure 7.4: Distribution of positron-neutron delay for the different event categories. The best fit curves are also drawn and the relative parameters indicated.

is an independent verification of the reliability of the adopted reconstruction technique.

7.2.2   Final selection

Both energy and topological cuts were studied and optimized by relying on Monte Carlo simulations of neutrino events. These predictions were cross-checked with the calibration data to gain confidence in the final efficiency numbers. The following criteria were finally adopted:

1) Positron energy: $E_{e+} < 8\,\mathrm{MeV}$:

2) Neutron energy: $6 < E_n < 12\,\mathrm{MeV}$:

3) Distance from geode boundary: $d_{e+} > 30\,\mathrm{cm}$, $d_n > 30\,\mathrm{cm}$:

4) Relative positron-neutron distance: $d_{e+n} < 100\,\mathrm{cm}$:

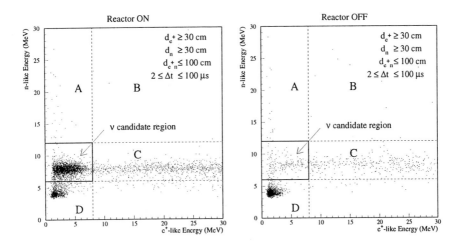

Figure 7.5: Neutron versus positron energy for neutrino-like events selected from the preliminary sample by applying the "topological" cuts here indicated.

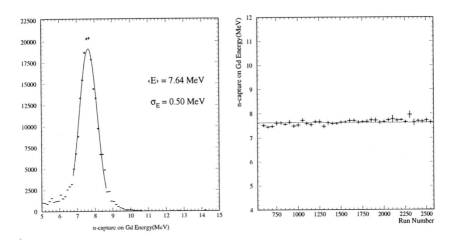

Figure 7.6: Neutron energy distribution for correlated background events (left) and average E_n vs. run number (right).

5) Neutron delay: $2 < \Delta t_{e+n} < 100\,\mu s$:

6) Neutron multiplicity: $N_n = 1$.

    The cut on the positron-like energy is chosen to accept all possible positron-like triggers. The L1lo trigger threshold limits the low end. The upper limit

was set to 8 MeV since the probability of having a larger positron energy
than this is negligible ($< 0.05\%$).

Cut 2) selects the neutrino events associated with the neutron capture on
gadolinium, which are more than 80% of the total. The lower cut introduces
an additional inefficiency due to γ rays escaping the containment region. As
illustrated in Fig. 7.5, 6 MeV is the most suitable choice to separate neutrino
events from the residual low energy uncorrelated background, thus optimizing
the signal to noise ratio.

The cut on the neutron delay covers about three capture times for neutrons in Region I. The 2 μs lower cut was introduced to reduce the effects of
the signal overshoot inherent in the AC coupling of PMT bases and front-
end electronics. These effects are particularly troublesome in the case of
NNADC's, which integrate current signals of either polarity.

The correlated background can be further reduced by applying a cut on
the secondary particle multiplicity[1]. As muon spallation processes usually
generate several neutrons, more than one particle is likely to enter the detector and give a detectable signal.

Events satisfying the selection criteria will be referred to as *neutrino
candidates* from now on. Let us now review the neutrino efficiency and the
background rejection operated by individual cuts.

7.2.3   Positron efficiency

Positron energy

The expected positron energy spectrum is obtained by folding the kinetic
energy spectrum coming out of (3.12) with the detector response function.
Two problems then arise in evaluating the positron efficiency. Firstly, the
neutrino spectral shape slightly varies along a reactor cycle as a consequence
of the fuel burn-up; however, the information daily provided by E.d.F. allows
us to accurately follow this variation. Secondly, the energy cut operated by
the L1lo trigger is increasing with time because of the aging of Region I
scintillator. To account for this effect also requires a daily check of the
equivalent energy threshold.

The procedure we followed to measure such a threshold is shown in
Fig. 7.7. A Lecroy qVt recorded the energy spectra due to a ^{60}Co source
placed at the detector centre. Two spectra were taken, the first one with an
external trigger provided by the L1lo signal and the second with an internal trigger set on a lower threshold. Besides the 2.5 MeV "sum" line, either
spectrum exhibits a low energy tail due to the energy loss in the calibration

---

[1]The positron (or primary) signal is associated with the L1 trigger preceding the L2.
Every signal associated with or occurring after the L2 is referred to as secondary.

pipe. The ratio of the two spectra (after background subtraction) yields the L1lo trigger efficiency curve. The equivalent energy threshold results from fitting this curve by an integral Gaussian function.

The energy threshold varies with time as illustrated in Fig. 7.8. Its behaviour is rather well described by a linear function. Discontinuities arise in coincidence with the threshold resetting, which was needed to bring the L1lo threshold back to the optimal value.

More extensive ^{60}Co calibrations were periodically performed at different positions of the source along the central pipe, so as to scan the z dependence of the energy threshold. The results are shown in Fig. 7.9 at four different stages of data acquisition. The overall dependence on position of the energy threshold can also be predicted by a Monte Carlo simulation of the QSUM and NSUM behaviour. The attenuation length values are the same used in the event reconstruction. The agreement of the threshold measurements with expectations is always good, as evident from the same figure. We can therefore rely on Monte Carlo predictions to determine the energy threshold at each position in the detector, where the source cannot be located. An expression for the positron threshold as a function of time and position can be derived by combining the daily threshold measurements at the detector centre with Monte Carlo evaluations. An explicit form of this threshold is given in cylindrical coordinates (ρ, z, ϕ) by the following equation:

$$E_{thr}(t, \overrightarrow{x}) = E_{thr}(t, \overrightarrow{x} = 0) \times (1 - \alpha(t)\rho^2) \times (1 - \beta(t)z^2) \qquad (7.1)$$

where the energy threshold at centre $E_{thr}(t, \overrightarrow{x} = 0)$ is extracted by interpolating the daily measurements shown in Fig. 7.8 and $\alpha(t), \beta(t)$ are linear functions obtained by Monte Carlo.

Distance from the geode boundary

The positron loss due to the selection 3) (distance from the geode boundary) was evaluated by a Monte Carlo simulation of neutrino interactions. 10000 events were uniformly generated in a volume including a 10 cm wide shell surrounding the target, so as to take into account spill-in/spill-out effects. The estimated efficiency comes out to be nearly independent of the scintillator degradation; we can then consider it as a constant and get the average value

$$\varepsilon_{d_{e+}} = (99.86 \pm 0.1)\% \qquad (7.2)$$

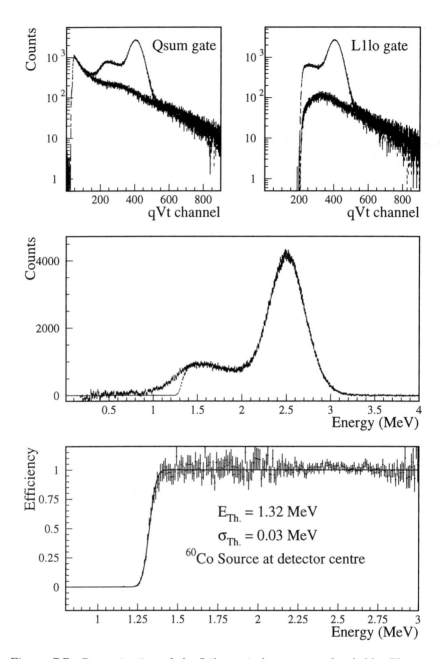

Figure 7.7: Determination of the L1lo equivalent energy threshold. The top figures show the QSUM spectra measured with a ^{60}Co at the detector centre by means of an internal and external triggered qVt (the corresponding background is superimposed). The central plot shows the background subtracted spectra. The bottom histogram, displaying the efficiency curve, follows an integral Gaussian function whose parameters are indicated.

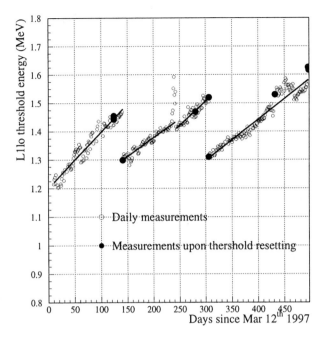

Figure 7.8: Equivalent energy threshold at detector centre as a function of time. The jumps visible here are due to the threshold setting.

7.2.4 Neutron efficiency

Neutron capture efficiency

The neutron capture efficiency is defined as the ratio of neutrons captured by gadolinium nuclei to the total number of captures. This ratio enters the global neutron efficiency since the neutron energy selection excludes events associated with neutron captures on hydrogen. This capture efficiency was studied by means of the standard ^{252}Cf source as well as with a special tagged ^{252}Cf source, providing a fission tagging signal. The calibration pipe was removed for these particular calibrations in order to avoid matter effects due to the iron content in the pipe; the source was housed in a plumb-line. Fission neutrons have a longer mean pathlength than neutrons from neutrino interactions (≈ 20 cm in scintillator instead of ≈ 6 cm). So calibration data must be coupled with Monte Carlo predictions to correct for effects associated with neutron spill-out (which are relevant for events approaching the boundary of Region I).

Neutron events are associated with Gd capture if $4 < E_n < 12$ MeV. The ratio obtained includes a 1% correction due to Gd-capture events with visible energy below 4 MeV. Combining data and MC information one obtains an

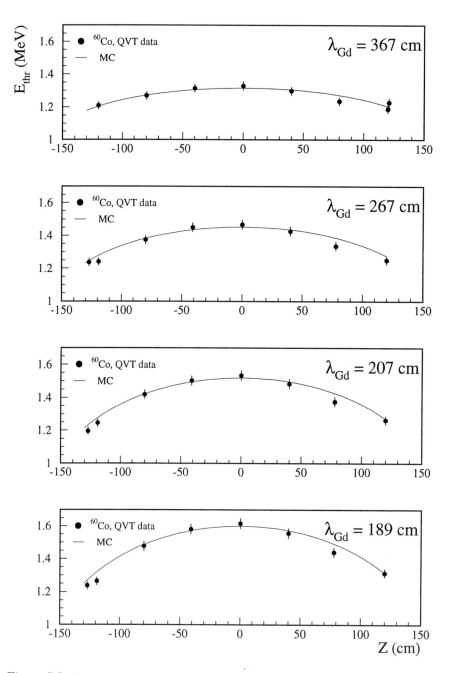

Figure 7.9: Energy threshold as a function of z for different values of the attenuation length for the Gd-doped scintillator. The measurements obtained with the ^{60}Co source follow the expected behaviour.

average efficiency over the entire detector

$$\varepsilon_{Gd} = (84.6 \pm 0.85)\% \qquad (7.3)$$

Neutron detection efficiency and e.m. energy containment

The 6 MeV lower energy cut introduces an inefficiency in neutron detection due to the γ rays from Gd-capture (by definition inside Region I) escaping the fiducial volume. Such an efficiency is almost independent of the neutron pathlength, so we can rely on the calibration data to evaluate this number; Monte Carlo predictions are used to cross-check the data. As this effect is expected to be more and more relevant for further and further outer events, a fine source scanning is required close to the target boudary. A special set of runs, besides the usual 40 cm-step scanning along the detector axis, is available at 2 cm steps, between 1.5 cm and 11.5 cm, from the bottom edge of the acrylic vessel.

The escape fraction is quoted by the ratio of the number of events with $4 < E_n < 6$ MeV over the ones with $4 < E_n < 12$ MeV. The values so-obtained range from 2.9% at the centre up to 6.6% at the bottom edge. By averaging over the target volume, one gets

$$\varepsilon_{E_n} = (94.6 \pm 0.4)\% \qquad (7.4)$$

Distance from the geode boundary

The procedure is the same followed to evaluate the corresponding positron efficiency. We obtain

$$\varepsilon_{d_n} = (99.5 \pm 0.1)\% \qquad (7.5)$$

Also this value is not significantly affected by scintillator degradation.

Delay cut efficiency

The neutron delay was studied by using both the ^{252}Cf and the Am/Be source; the latter emits one neutron per fission; this prevents biases, due to the ADC integration dead time, arising at high neutron multiplicities. Fig. 7.10 shows the delay distributions obtained with the Am/Be source placed at the centre and at the bottom edge of the target. The neutron capture time at the centre can be used to determine the Gd content in the Region I scintillator (to be used in the Monte Carlo code). The fitted decay time $\tau = (30.7 \pm 0.5)\,\mu s$ corresponds to a concentration of $(0.940 \pm 0.005)\%_o$ in mass. Given the chemical composition of the scintillator, it is possible to rely on the Monte Carlo simulation to predict the efficiency related to the neutron delay cut.

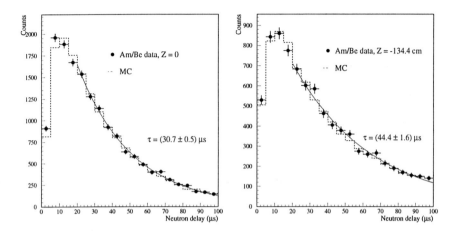

Figure 7.10: Neutron delay distribution measured with the Am/Be source at the detector centre (left) and at the bottom edge of the acrylic vessel (right).

The estimated loss due to the $2\,\mu s$ cut amounts to $1.6 \pm 0.2\%$; the fraction of neutrons with $\Delta t > 100\,\mu s$ is $4.7 \pm 0.3\%$. We can then conclude

$$\varepsilon_{\Delta t} = (93.7 \pm 0.4)\% \qquad (7.6)$$

7.2.5   Distance cut efficiency

We again used the reconstruction of 10000 Monte Carlo generated events to predict the efficiency due to the positron-neutron distance cut. We obtained

$$\varepsilon_{d_{e+n}} = (98.4 \pm 0.3)\% \qquad (7.7)$$

Also this efficiency is nearly independent of the time evolution of the Gd-doped scintillator.

7.2.6   Neutron multiplicity

The neutron multiplicity cut rejects neutrino events if a "spurious" L1lo trigger is superimposed on the positron-neutron pair. L1lo triggers are mainly associated with γ-background events, $\approx 97\%$ of which have energies lower than the high threshold; so, the background above the L1hi threshold can be neglected in what follows. The error introduced by this approximation is negligible.

Let us consider all possible sequences:

1) $e^+ - n - \gamma$ with $t_\gamma - t_n < 100\,\mu s$:

2) $e^+ - \gamma - n$:

3) $\gamma - e^+ - n$ with $t_{e^+} - t_\gamma < 100\,\mu s$:

Case 1

The L2 triggers on the (e^+, n) pair and the γ signal is mistaken for a second neutron; the neutrino event is then rejected because of cut 6). Given the γ rate R_γ, the probability of such a sequence is

$$1 - \varepsilon_1 = R_\gamma \Delta t_\nu \tag{7.8}$$

from which ε_1 is extracted. The average efficiency is $(98.6 \pm 0.3)\%$.

Case 2

If the positron energy exceeds the high threshold, the L2 triggers on the (e^+, γ) pair; then two neutron-like signals are detected (the "true" neutron plus the γ) and the neutrino event is rejected by cut 6). If the positron energy is lower then the L1hi threshold, the L2 triggers on the (γ, n) and the γ is mistaken for a positron-like event. Since the γ signal is not correlated to the neutron, the probability of surviving the topological cuts 3) and 4) is quite low ($< 2\%$) and can be neglected. As a conclusion, neutrino events occurring in such a sequence are always rejected, whatever the positron energy is.

Let $P_n(t_c)$ be the neutron delay distribution shown in Fig. 7.10 and t the time between the positron and the γ signals. The neutrino efficiency can be written as follows:

$$\varepsilon_2 = 1 - \int_0^{100} P_n(t_c)\, dt_c \int_0^{t_c} R_\gamma\, dt = 1 - R_\gamma \overline{t_c} \tag{7.9}$$

where $\overline{t_c} = (30.5 \pm 0.5)\,\mu s$ is the average neutron delay.

Case 3

As in the previous case, if the positron signal fulfilled the high trigger condition, the L2 would trigger on the (γ, e^+) pair and the selection criterium 6) would reject the event; otherwise the L2 would trigger on the (e^+, n) pair and the neutrino events would be accepted. So

$$\varepsilon_3 = 1 - R_\gamma f_H \Delta t_\nu \tag{7.10}$$

$f_H = (0.45 \pm 0.05)$ being the positron fraction above the high energy threshold.

Combined efficiency

The joint neutron multiplicity efficiency is obtained by multiplying the three above expressions. It must be evaluated on a run to run basis, as the R_γ depends on the time behaviour of the L1lo threshold. The average value is $\varepsilon_{2n} = (97.4 \pm 0.5)\%$.

Table 7.2: Summary of the neutrino detection efficiencies.

selection	efficiency (%)	relative error (%)
positron energy*	97.8	0.8
positron-geode distance	99.9	0.1
neutron capture	84.6	1.0
capture energy containment	94.6	0.4
neutron-geode distance	99.5	0.1
neutron delay	93.7	0.4
positron-neutron distance	98.4	0.3
neutron multiplicity*	97.4	0.5
combined*	69.8	1.5

*average values

7.3   The neutrino signal

In Figs. 7.11 through 7.13 the final neutrino candidate sample, with all the selection cuts applied, is presented. A total number of 2991 neutrino candidates
was collected, 287 of which during the reactor-off periods. For proper comparison with expectations, the entire acquisition cycle was divided according
to the dates of the threshold resetting, when variations in the background
rate were expected. For each resulting period, the reactor-off background
was normalized to the same livetime as the reactor-on spectra.

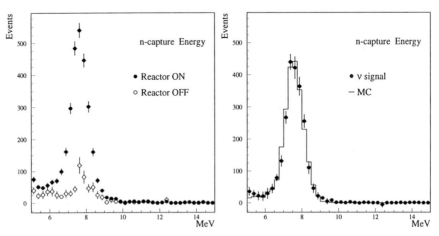

Figure 7.11: Neutron energy spectra for reactor-on and reactor-off periods (left)
and background subtracted spectrum compared to Monte Carlo expectations
(right).

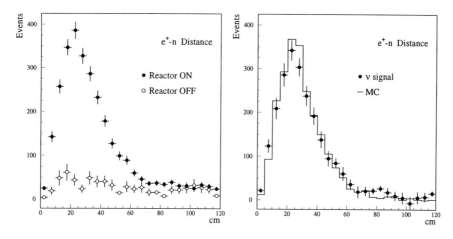

Figure 7.12: Same as before, for the positron-neutron distance.

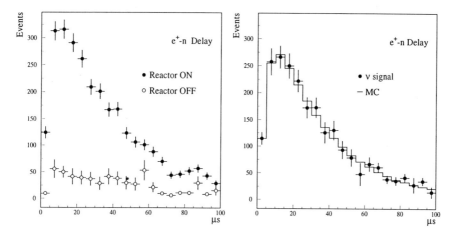

Figure 7.13: Neutron delay distributions for reactor-on and reactor-off periods (left) and background subtracted spectrum compared to MC predictions (right).

7.4 The positron spectrum

7.4.1 Measured spectrum

Fig. 7.14 shows the complete measured spectra (reactor-on, reactor-off), obtained by summing the spectra collected during runs relative to different off-line periods; the resulting positron spectrum (reactor-on minus reactor-off) is presented in Fig. 7.15. The chosen bin width (0.4 MeV) is roughly adapted both to statistics and to energy resolution.

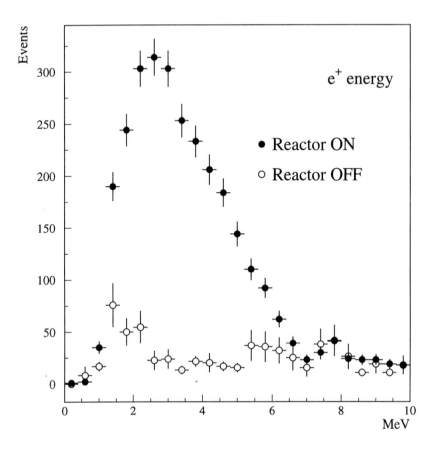

Figure 7.14: Experimental positron spectra for reactor-on and reactor-off periods after application of all selection criteria. The errors shown are statistical.

7.4.2   Predicted spectrum

The expected visible energy positron spectrum at the detector position, for a mean reactor-detector distance L_k, is given by:

$$S_k(E, L_k, \theta, \delta m^2) = \frac{1}{4\pi L_k^2} n_p \int h(L, L_k)\, \mathrm{d}L \int \sigma(E_{e^+}) S_\nu(E_\nu) \times$$
$$P(E_\nu, L, \theta, \delta m^2) r(E_{e^+}, E)\varepsilon(E_{e^+})\, \mathrm{d}E_{e^+}, \qquad (7.11)$$

where

E_ν, E_{e^+}	are related by (3.12),
n_p	is the total number of target protons in the Region I scintillator,
$\sigma(E_{e^+})$	is the detection cross section (3.13),
$S_\nu(E_\nu)$	is the composite antineutrino spectrum,

$h(L, L_k)$	is the spatial distribution function for the finite core and detector sizes,
$r(E_{e^+}, E)$	is the detector response function providing the visible positron energy,
$P(E_\nu, L, \theta, \delta m^2)$	is the two-flavour survival probability,
$\varepsilon(E_{e^+})$	is the combined detection efficiency.

We first calculated the positron spectrum in the absence of neutrino oscillations, by using the Monte Carlo code to simulate both reactors and the detector. The composite antineutrino spectrum was generated for each of the 205 fuel elements of each reactor core; each antineutrino was assigned a weight according to the prescriptions given at §3.8. The interaction points were randomly chosen in the target; the positron and resulting annihilation photons were tracked in the detector, and scintillator saturation effects were included to correctly evaluate the positron visible energy. The resulting spectrum, summed over the two reactors, is superimposed on the one measured in Fig. 7.15 to emphasize the agreement of the data with the no-oscillation hypothesis; the Kolmogorov-Smirnov test for compatibility of the two distributions gives a 82% probability. The measured vs. expected ratio, averaged over the energy spectrum (also presented in Fig. 7.15) is

$$R = 1.01 \pm 2.8\%(\text{stat}) \pm 2.7\%(\text{syst}) \tag{7.12}$$

In the next Chapter we will discuss how this ratio constrains the oscillation hypothesis.

7.5 The background

One of the main values of the experiment, the low background, is in fact a problem when measuring its induced rate for background subtraction. Apart from the low statistics, one further difficulty is that the background rate depends on the trigger conditions which, as discussed above, changed with time because of the scintillator aging and of the positron threshold adjustments. Separate estimates of the background are then needed for each data taking period. Only 34 events were collected during the 1997 run (until January 12th 1998, date of the last threshold resetting) with reactors off, the total live time being 577 h; this implies a background rate of 1.41 ± 0.24 events per day. Most of the reactor-off statistics, amounting to 253 events, was collected during the 1998 run when the cumulative reactor power was very low (see

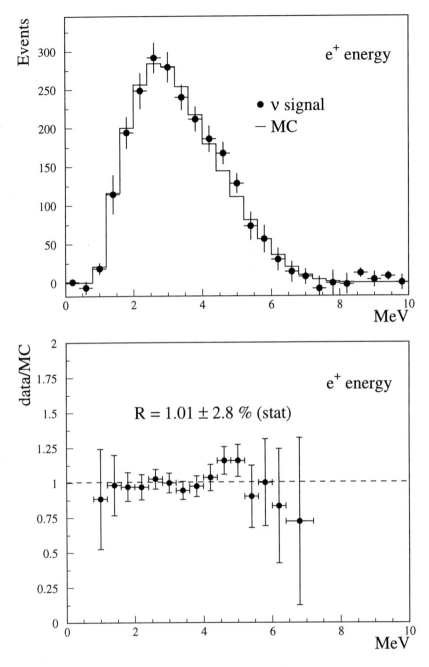

Figure 7.15: (above) Expected positron spectrum for the case of no oscillations, superimposed on the measured positron spectrum obtained from the subtraction of reactor-on and reactor-off spectra; (below) measured over expected ratio. The errors shown are statistical.

Tab. 7.4 below); the reactor-off live time is 2737 h. The resulting background rate is 2.22 ± 0.14, about twice as large as in the 1997 run.

A possible explanation for this variation relies on the lowering of the NSUM threshold associated with the L1lo trigger. A lower number of hit PMT's implies a larger fiducial volume extended towards the PMT boundary, where the event rate is dominated by the natural radioactivity; moreover, we saw that the reconstruction algorithm overestimates the energy of events approaching the PMT's (see Fig. 6.15). Therefore a larger fraction of radioactivity events drifts towards the neutron energy window, that is to say, from region D to the neutrino candidate window with reference to Figs. 7.2,7.3,7.5; as a result, the accidental component of the background is largely enhanced. The correlated background instead is nearly unaffected by the low trigger conditions, since the neutron signal is much higher than the threshold. We verified that the correlated background does not significantly change throughout the experiment.

As a further cross-check of the reactor-off estimates, we measured the background by extrapolating to zero the candidate yield versus reactor power. This will be the subject of §7.6.

7.5.1 Correlated background

Fig. 7.16 shows the energy distribution of e^+-like signals associated with the correlated background; events are selected by applying all the criteria for neutrino candidate selection, except the e^+-like energy. Apart from a low-energy peak (mostly due to the pile-up of neutrino events and accidental background), the spectrum exhibits a roughly flat trend extending beyond 30 MeV, with a slight increase at higher energies due to the NNADC saturation. A Monte Carlo simulation of this spectrum is very difficult, since no reliable transport code is available for spallation neutrons in the $10 \div 100$ MeV kinetic energy range; moreover, the observed spectrum is affected by the scintillator saturation (which is relevant at low recoil proton energies).

In order to verify the stability of the correlated background rate, we divided the complete run into three different periods (see Tab. 7.4 below) according to the dates of the threshold resetting; the spectra for each period were fitted by a constant in the $10 < E < 30$ MeV range, so obtaining respectively $(11.9 \pm 0.8)\,\mathrm{MeV}^{-1}$, $(19.4 \pm 1.0)\,\mathrm{MeV}^{-1}$, $(20.6 \pm 1.0)\,\mathrm{MeV}^{-1}$. Dividing by the live times (also listed in Tab. 7.4), we finally found

$$B_{corr} = \begin{cases} (0.156 \pm 0.01)\,\mathrm{MeV}^{-1}\,\mathrm{d}^{-1} & \text{for the 1st period,} \\ (0.158 \pm 0.01)\,\mathrm{MeV}^{-1}\,\mathrm{d}^{-1} & \text{for the 2nd period,} \\ (0.151 \pm 0.01)\,\mathrm{MeV}^{-1}\,\mathrm{d}^{-1} & \text{for the 3rd period,} \end{cases} \qquad (7.13)$$

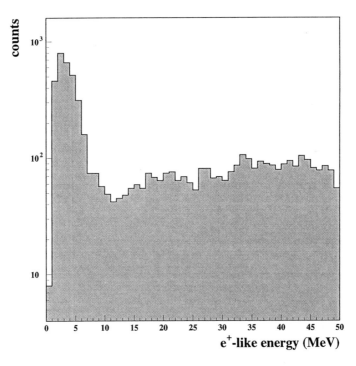

Figure 7.16: Energy distribution of e^+-like signals associated with the correlated background.

which confirms the stability of the correlated background rate throughout the experiment.

A raw evaluation of this rate is obtained by assuming a flat energy spectrum also in the positron window ($E_{thr} < E < 8\,\text{MeV}$). Taking the average value from (7.13) and assuming $E_{thr} = 1.5\,\text{MeV}$, we obtained a rate of $(1.01 \pm 0.04(stat) \pm 0.1(syst))\text{events d}^{-1}$.

7.5.2   Accidental background

The accidental background rate was determined by separate estimates of the single rate for both e^+-like and n-like signals. In order to minimize possible biases due to the trigger, both estimations were performed by looking at the "isolated old" events, i.e. at the hits recorded in the NNADC event buffer (storing up to 9 events) at least 2 ms before the L2 trigger and occurring at least 1 ms after the preceeding and before the following signal. The event selection was operated by applying the same cuts (energy and distance from edge) used for the candidate selection; in the 1997 run we found $R_{e^+} = (64.8 \pm 0.1)\,\text{s}^{-1}$ for the e^+-like event rate and $R_n = (45 \pm 2)\,\text{h}^{-1}$ for the n-like one. The rate of accidental coincidences $e^+ - n$ in the $2 \div 100\,\mu s$ time window was then $R_{e^+} \times R_n \times 98\,\mu s = (7.0 \pm 0.3)\,d^{-1}$. In order to obtain the contribution of these accidentals to the candidate rate, we had to divide the last number by 2 since we were looking for only $e^+ - n$ sequences, whereas the given rate contains either sequence. An additional reduction factor $f_d = (0.12 \pm 0.01)$ was applied to account for the selection operated by the $e^+ - n$ distance cut. The resulting background rate was $(0.42 \pm 0.05)\,d^{-1}$, which is in agreement with the previous determinations of the total and the correlated background for the 1997 run.

7.6 Neutrino yield versus power

The usual method in reactor experiments to measure the neutrino deficit is to determine the ratio of measured vs. predicted neutrinos via a spectral shape comparison. However, the Chooz experiment had the unique opportunity to start measuring the neutrino flux before either reactor started working and also after the reactor shutdown. This produced two advantages: first, the possibility to collect enough reactor-off data to determine the background rate, as we did in the previous section; second, the possibility to measure the neutrino flux while the reactors were ramping up to full power. By fitting the slope of the measured rate versus reactor power, one determines the neutrino rate at full power, to be compared with the predicted neutrino production rate at full power. This provides a very powerful tool to determine the neutrino deficit and test the oscillation hypothesis. Moreover, one obtains an independent estimate of the background rate to be compared with the values quoted above.

The fitting procedure proceeds as follows. For each run the predicted number of neutrino candidates results from the sum of a signal term, linearly depending on the reactor power, and the background, assumed to be constant and independent of power; so,

$$\overline{N}_i = (B + W_{1i}Y_{1i} + W_{2i}Y_{2i})\Delta t_i, \qquad (7.14)$$

where the index i labels the run number, Δt_i is the corresponding live time, B is the background rate and (Y_{1i}, Y_{2i}) the positron yields induced by each reactor. These yields still depend on the reactor index (even in the absence of neutrino oscillations), because of the different distances, and on run number, as a consequence of their different fissile isotope composition. It is thus convenient to factorize Y_{ki} into a function X_k (common to both reactors in the no-oscillation case) and distance dependent terms, as follows:

$$Y_{ki} = (1 + \eta_{ki})\frac{L_1^2}{L_k^2}X_k, \qquad (7.15)$$

where $k = 1, 2$ labels the reactors and the η_{ki} corrections contain the dependence of the neutrino counting rate on the fissile isotope composition of the reactor core. We are thus led to define a cumulative "effective" power according to the expression [2]

$$W_i^* \equiv \sum_{k=1}^{2} W_{ki}(1 + \eta_{ki})\frac{L_1^2}{L_k^2};$$ (7.16)

eq.(7.14) then becomes

$$\overline{N}_i = (B + W_i^* X)\Delta t_i,$$ (7.17)

X being the positron yield per unit power averaged over the two reactors.

The burn-up corrections η_{ki} must be evaluated on a run by run basis. Since using the GEANT routines would have been very time-consuming for such a job, we preferred to follow an approach slightly different from that explained at §7.4. 10000 $\overline{\nu}_e$'s per run were generated at each reactor according to the respective fuel composition; the kinetic positron energy was obtained by the simple relation $E_\nu = T_{e+} + 1.804\,\text{MeV}$ (which comes from (3.12) in the limit of infinite nucleon mass). The cross section (3.13) was thus multiplied by a function $\delta(E_\nu)$ to correct for the shift in the positron energy scale due to the finite neutron recoil effect [93]; this correction can be parametrized as

$$\delta(E_\nu) = 1 - 0.155 \exp\left(\frac{E_\nu/\,\text{MeV} - 8}{1.4}\right)$$ (7.18)

We then applied the detector response function (evaluated by Monte Carlo simulations at several positron energies ranging from 0.5 MeV up to 10 MeV) to the positron kinetic energy and weighted each event according to the positron threshold efficiency.

We built the likelihood function $\mathcal{L}$ by the joint Poissonian probability of detecting N_i neutrino candidates when $\overline{N}_i$ are expected, and defined

$$F \equiv -\ln \mathcal{L} = -\sum_{i=1}^{n} \ln P(N_i; \overline{N}_i)$$ (7.19)

Searching for the maximum likelihood to determine the parameters X and B is then equivalent to minimizing (7.19). The minimization procedure is similar to that used in the event reconstruction (see §6.6); a first minimization step calls the SIMPLEX minimizer to have a first estimation of the fit

---

[2]The "effective" power may be intended as the thermal power released by a one-reactor station located at the reactor 1 site, providing 9.55 GW at full operating conditions and at starting of reactor operation.

parameters; the SIMPLEX output parameters are then used to initialize MI-GRAD, which computes the best-fit parameters as well as the error matrix elements. After this minimization step, the Poissonian χ^2 associated with the likelihood function (defined by (6.9)) is evaluated in order to test the goodness of fit.

Both the average positron yield X and the background rate B are assumed to be constant. This is true, by definition, for the positron yield, since the effect of the threshold resetting on the positron efficiency is already included in the η_{ki} correction factors. On the contrary, we saw that significant variations of the background rate are expected as soon as the trigger thresholds are reset. So we are again forced to divide the complete run sample into three periods, according to the dates of the threshold resetting, and draw the fit parameters for each period separately. The results are listed in Tab. 7.4.

Table 7.4: Summary of the likelihood fit parameters for the three data taking periods.

period	1	2	3
starting date	97/4/7	97/7/30	98/1/12
runs	579 → 1074	1082 → 1775	1778 → 2567
live time (h)	1831.3	2938.8	3268.4
reactor-off time (h)	38.9	539.5	2737.2
$\int W \, dt$ (GWh)	7798	10636	2838
B (counts d^{-1})	1.25 ± 0.6	1.22 ± 0.21	2.2 ± 0.14
X (counts d^{-1} GW^{-1})	2.60 ± 0.17	2.60 ± 0.09	2.51 ± 0.17
χ^2/dof	136/117	135/154	168/184
N_ν (counts d^{-1}) (@@full power)	24.8 ± 1.6	24.8 ± 0.9	24.0 ± 1.6

The best neutrino signal estimation comes from period 2, when the neutrino integrated luminosity reached its maximum value. The cumulated thermal energy (proportional to the neutrino flux) during the first period is $\approx 20\%$ lower than that; moreover, the reactor-off time is much lower, thus affecting the background measurement as well as the neutrino yield estimation (whose uncertainty is twice as large as in period 2). Period 3 is the one with the lowest neutrino statistics; nevertheless, the cumulated reactor-off time is by far the largest, so that the overall uncertainty on the neutrino yield is similar to period 1.

By averaging the signal X over the three periods, one obtains

$$\langle X \rangle = (2.58 \pm 0.07) \text{ counts d}^{-1}\,\text{GW}^{-1}, \qquad (7.20)$$

corresponding to (24.7 ± 0.07) daily neutrino interactions at full power; the overall statistical uncertainty then amounts to 2.8%.

7.6.1 Neutrino yield for individual reactors

The same fitting procedure can be extended to determine the contribution to the neutrino yield from each reactor and for each energy bin of the positron spectra. After splitting the signal term into separate yields and introducing a dependence on the positron energy, eq.(7.17) can be rewritten in the form

$$\overline{N}_i(E_j) = (B(E_j) + W_{1i}^*(E_j)X_1(E_j) + W_{2i}^*(E_j)X_2(E_j))\Delta t_i \qquad (7.21)$$

The spectrum shape is expected to vary, due to fuel aging, throughout the reactor cycle. Burn-up correction factors η_{ki} then need to be calculated for each bin of the positron spectrum. The fitted yields, averaged over the three periods, are listed in Tab. 7.5 and plotted in Fig. 7.17 against the expected yield in the absence of neutrino oscillations. The yield parameters X_1, X_2 are

Table 7.5: Experimental positron yields for both reactors (X_1 and X_2)and expected spectrum ($\tilde{X}$) for no oscillation. The errors (68% C.L.) and the covariance matrix off-diagonal elements are also listed.

E_{e^+} (MeV)	$X_1 \pm \sigma_1$	$X_2 \pm \sigma_2$ (counts d^{-1} GW^{-1})	$\tilde{X}$	σ_{12} (counts d^{-1} GW^{-1})^2
1.2	0.151 ± 0.031	0.176 ± 0.035	0.172	$-2.2 \cdot 10^{-4}$
2.0	0.490 ± 0.039	0.510 ± 0.047	0.532	$-1.5 \cdot 10^{-4}$
2.8	0.656 ± 0.041	0.610 ± 0.049	0.632	$-3.5 \cdot 10^{-4}$
3.6	0.515 ± 0.036	0.528 ± 0.044	0.530	$-3.3 \cdot 10^{-4}$
4.4	0.412 ± 0.033	0.408 ± 0.040	0.379	$-2.0 \cdot 10^{-4}$
5.2	0.248 ± 0.030	0.231 ± 0.034	0.208	$-0.7 \cdot 10^{-4}$
6.0	0.102 ± 0.023	0.085 ± 0.026	0.101	$-1.3 \cdot 10^{-4}$

slightly correlated, as shown in Tab. 7.5; such a correlation (which does not exceed 20%) is always negative, since, at given candidate and background rates, an increase in reactor 1 yield corresponds to a decrease in reactor 2 yield (and *vice versa*). In the next Chapter, while building the right χ^2 statistic to test the oscillation hypothesis, we shall take the covariance matrix into account.

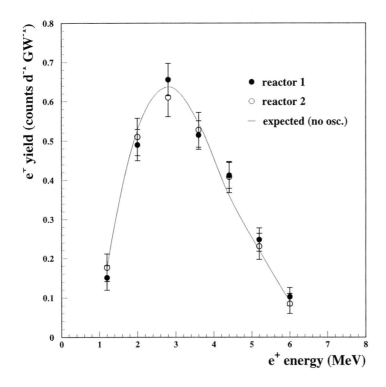

Figure 7.17: Positron yields for the two reactors, as compared with expected yield for no oscillations.

7.6.2   Neutrino yield versus fuel burn-up

We have repeatedly stated that the contributions of the main fissile iso-topes to the thermal power change in the course of an operating period; this should produce a corresponding decrease in the total neutrino counting rate as well as a modification in the spectral shape. The magnitude of this variation amounts to $\approx 10\%$ throughout the first cycle of the Chooz reactors, as we have already shown in Chapter 3, which by far exceeds the statistical and systematic accuracy of the neutrino flux. We were then forced to follow the dynamics of the fuel burn-up and to apply daily corrections to the thermal power in order to restore the linearity with the positron yield (see eqs.(7.15),(7.16)).

We tried to see whether the neutrino counting rate varies with the reactor burn-up according to predictions, with the purpose of improving the reliability and internal consistency of our results. The number N_ν of neutrino

events, recorded in a time interval Δt, by a detector at a distance L from
the core of a reactor working at a power W can be derived from eq.(3.9); by
inverting we find

$$\frac{\sigma_f}{E_f} = 4\pi \frac{L^2}{W \Delta t} \frac{N_\nu}{n_p \varepsilon} \qquad (7.22)$$

σ_f and E_f being respectively the reaction cross section (3.13) and the average energy absorbed in the core in a single fission. The ratio σ_f/E_f, which
contains the dependence on the fuel composition, can be evaluated for either reactor by considering the runs where reactors were alternatively on;
runs are selected if the thermal energy is $\int W \, dt > 850 \, \text{GW d}$ (which is the
energy released in one day by a reactor at 20% of full power). The num-
ber of neutrino events is obtained by subtracting the number of background
events (as determined in the previous Section) from the number of candidate
events collected in a run. The resulting σ_f/E_f values are finally grouped in
1000 MW d burn-up intervals and plotted vs. reactor burn-up in Fig. 7.18.

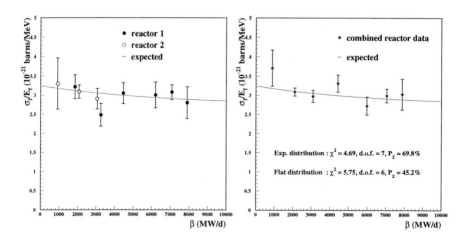

Figure 7.18: Variation of the measured neutrino counting rate, as a function of the
fuel burn-up for separate (left) and combined (right) reactor data and comparison
with predictions. Error bars include only statistical uncertainties.

A more significant test is performed by using combined reactor information. Each run is assigned an average burn-up defined as

$$\overline{\beta} \equiv \frac{\beta_1 W_1/L_1^2 + \beta_2 W_2/L_2^2}{W_1/L_1^2 + W_2/L_2^2}; \qquad (7.23)$$

similarly, σ_f/E_f is obtained by

$$\frac{\sigma_f}{E_f} = 4\pi\left(\frac{L_1^2}{W_1\Delta t} + \frac{L_2^2}{W_2\Delta t}\right)\frac{N_\nu}{n_p\varepsilon} \tag{7.24}$$

The resulting values are plotted in Fig. 7.18 and compared with the expectations; the agreement is excellent, since $\chi^2 = 4.69$ with 7 d.o.f., corresponding to a 69.8% χ^2-probability. The compatibility of the data with a flat distribution is still good ($\chi^2 = 5.75$ with 6 d.o.f., $P_{\chi^2} = 45.2\%$), but lower than predictions.

7.7   Neutrino direction

The use of reaction (2.22) to detect low energy antineutrinos in large volume scintillator detectors, is the best tool to measure the antineutrino energy spectrum; moreover, it provides a good determination of the antineutrino incoming direction. This determination is based on the neutron boost in the forward direction, as a result of the kinematics of the above reaction; the neutron then retains a memory of the source direction, which survives even after collisions with the protons in the moderating medium.

In Chooz we exploited this neutron recoil technique to locate the reactor direction, with the twofold objective of testing our event reconstruction method and tuning our Monte Carlo simulations of the slowing down and capture of neutrons in the scintillator. We then studied a possible extension of this technique to much larger scintillation detectors, such as Kamland, and their capabilities in locating astrophysical neutrino sources, such as Supernovæ.

7.7.1   Location of the reactors

The positron angular distribution for the reaction (2.22), for low energy antineutrinos, can be written in the form [113]

$$\frac{d\sigma}{d\cos\theta} \simeq 1 + v_{e^+}a_0\cos\theta, \tag{7.25}$$

where θ is the angle between the antineutrino and the positron directions in the laboratory frame, and v_{e^+} is the positron velocity (in $c = 1$ units); the amplitude a_0 depends on the vector and axial weak form factors ($f = 1$, $g = 1.26$ in the $q^2 \to 0$ limit) by the following expression

$$a_0 = \frac{f^2 - g^2}{f^2 + 3g^2} \simeq -0.1 \tag{7.26}$$

This weak backward asymmetry could be used to recognize the incoming neutrino direction, provided that a huge number of events is collected. In the case of Chooz, by averaging the cross section over the solid angle and the positron energy spectrum, one obtains a displacement of the positron from the interaction vertex ≈ -0.05 cm; as $\sigma_x \simeq 15$ cm for the positron, ≈ 100000 events would be needed to significantly observe such a tiny anisotropy.

On the contrary, the neutron emission angle with respect to the incident neutrino direction is limited to values below $\sim 55°$; this results from Fig. 7.19, where the neutron angle θ_n is plotted vs. the neutron kinetic energy T_n (which extends up to ~ 100 KeV). Moreover, the neutron moderation maintains some memory of the initial neutron direction [114]; as a matter of fact, in each elastic scattering the average cosine of the outgoing neutron is $\langle \cos \theta_n \rangle = 2/3A$, A being the mass number of the scattering nucleus. The direction is thus best preserved by collisions on protons, which is also the most effective target nucleus at energies below 1 MeV. The neutron mean free path rapidly reduces during moderation, since the scattering cross section rapidly increases at lower and lower neutron energies; so the bulk of the neutron displacement is due to the first two or three collisions. The isotropic diffusion of thermalized neutrons does not affect the average neutron displacement along the neutrino direction.

The average neutron displacement in Chooz is evaluated to be 1.7 cm. Since the experimental position resolution is $\sigma_x \approx 19$ cm for the neutron and the collected neutrino statistic is ≈ 2500, the precision of the method is ≈ 0.4 cm; so the neutron displacement can be observed at $\sim 4\sigma$ level. The average direction (in spherical coordinates) of the two reactors in the Chooz detector frame was measured by standard surveying techniques to be $\phi = (-50.3 \pm 0.5)°$ ($\hat{x}$ being the polar axis) and $\theta = (91.5 \pm 0.5)°$.

In order to measure the direction of this average displacement, we define for each neutrino candidate the unit vector with its origin in the positron and pointing to the neutron capture reconstructed position. Besides the standard selection criteria, we required the combined thermal power of both reactors to be ≥ 3 GW, so as to enhance the signal to noise ratio; this reduces the candidates to ≈ 2500. The distribution of the projection of this unit vector along the known direction over the selected candidate sample is shown in Fig. 7.20; this is compared with the expected distribution from a Monte Carlo simulation with a higher statistic (5000 events). Both distributions present an evident forward asymmetry although their r.m.s. (0.570 and 0.565 respectively) are quite close to that of a flat distribution.

We verified that the same technique, when applied to ^{252}Cf runs (where the prompt fission γ's fake the positron signal), gives results compatible with isotropic distributions. This excludes that the observed anisotropy could be

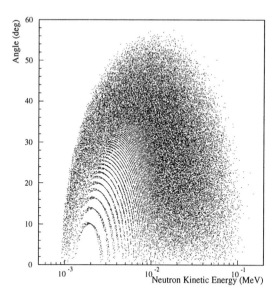

Figure 7.19: Neutron emission angle (with respect to the incident $\overline{\nu}_e$ direction) vs. its kinetic energy; the discrete structure of lower-left part of the picture is an effect of the logarithmic scale for the abscissa combined with the $\overline{\nu}_e$ energy binning.

due to possible biases in the event reconstruction.

Our determination of the neutrino direction comes from averaging the unit positron-neutron vector over the selected candidate sample; so

$$\vec{p} = \frac{1}{N} \sum_i \frac{\vec{x}^i_n - \vec{x}^i_{e+}}{|\vec{x}^i_n - \vec{x}^i_{e+}|} \tag{7.27}$$

The uncertainty δ on that direction can be quoted as the half-width of the cone around $\vec{p}$ containing 68% of the integral of the $\vec{p}$ distribution. This distribution is known *a priori*, since $\vec{p}$ is the sum (divided by N) of N variables whose average ($\vec{p}$ itself) and r.m.s. (assumed to be $\sigma = 1/\sqrt{3}$ for each coordinate) are known; the central limit theorem then states that each of the three $\vec{p}$ components is Gaussian-distributed around the average value with $\sigma = 1/\sqrt{3N}$. The neutrino direction resulting from the data, with the relative uncertainty, is presented in Tab. 7.6; the Monte Carlo predictions, for a sample with the same statistics, is also listed for comparison. The measured direction has 16% probability to be compatible with that expected, whereas the probability of being a fluctuation of an isotropic distribution is negligible.

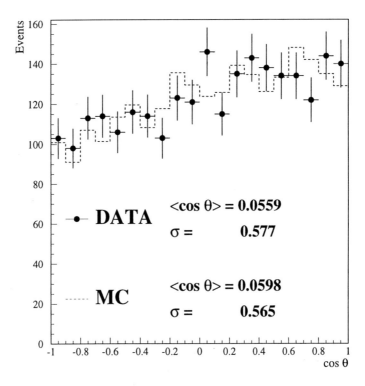

Figure 7.20: Distribution of the projection of the positron-neutron unit vector along the incident direction for the selected neutrino sample; the data are compared with a higher statistic (5000 events) Monte Carlo distribution.

Table 7.6: Measurement of neutrino direction: data and Monte Carlo.

	$|\vec{p}|$	ϕ	θ	δ
Data	0.055	$-70°$	$103°$	$18°$
MC	0.052	$-56°$	$100°$	$19°$

The average neutron displacement is estimated to be 1.9 ± 0.4, in agreement with the expected value.

7.7.2   Implication for SN neutrinos

The above technique was then trained to determine the neutrino direction from a Supernova in a liquid scintillator experiment. The harder neutrino spectrum (the average detected neutrino energy is $\approx 17\,\mathrm{MeV}$ is such a case) has two differing effects to the Supernova localization capability:

- the maximum neutron angle with respect to the neutrino direction increases with the neutrino energy ($\cos(\theta_n)_{max} \propto 1/\sqrt{E_\nu}$ in the infinite nucleon mass limit)

- higher neutron energies imply lower scattering cross–sections and, as a consequence, larger displacements of the neutron capture point from the neutrino interaction vertex.

The combination of these two effects was evaluated by using the full Chooz Monte Carlo simulation; the Supernova $\overline{\nu}_e$ spectrum was described by a pseudo Fermi-Dirac distribution given by [115]

$$\frac{\mathrm{d}N}{\mathrm{d}E_\nu} = C\frac{E_\nu^2}{1 + \exp(E_\nu/T)} \tag{7.28}$$

with temperature $T = 3.3\,\mathrm{MeV}^3$. We generated 5000 neutrino interactions in an experiment with the same geometry, the same position resolution and the same target (Gd-loaded liquid scintillator) as Chooz; for instance, this number of events could be detected in a liquid scintillator experiment with mass equal to that of SuperKamiokande. The neutrino direction was chosen to be at the zenith ($\theta = 0$).

Table 7.7: Determination of the Supernova neutrino direction, as obtained from Monte Carlo events. The results in the second row are obtained by requiring the positron–neutron distance to be larger than 20 cm. Note that the ϕ angle determination is irrelevant since neutrinos are directed along the zenith axis.

$|\vec{p}|$	ϕ	θ	δ
0.079	111°	11°	8.8°
0.102	66°	8°	8.4°

The results obtained by applying the above technique to this case are listed in Tab. 7.7. The average unit positron-neutron vector has a larger component along the neutrino direction than in the reactor case; it follows that the effect of a larger neutron displacement dominates over the broadening of the neutron emission angle. We also investigated the possibility of improving the direction uncertainty by applying a selection on the positron-neutron distance. The best situation arises upon requiring this distance to be larger than 20 cm, which corresponds to a reduction of the neutrino sample from 5000 to ~ 3400; the results of this selection are also shown in Tab. 7.7.

---

[3]Other authors[113] suggest using more energetic spectra, which (as we shall see shortly) improve the possibility of determinining the $\overline{\nu}_e$ direction with the present method.

Chapter 8

Neutrino oscillation tests

The neutrino data sample presented in the previous chapter allows us to extract the confidence domain in the plane of the oscillation parameters $(\sin^2(2\theta), \delta m^2)$. Since no evidence was found for a deficit of measured vs. expected neutrinos, the no-oscillation hypothesis must be included in that domain; the oscillation contour will thus result in an exclusion plot, where the allowed region for oscillation is the area to the left of and below the contour line.

We followed three different approaches for analyzing the results of the experiment. The first one ("analysis A" in what follows) uses the predicted positron spectrum (obtained by merging the reactor information, neutrino spectrum models and detector response) in addition to the measured spectra for each reactor. This approach introduces the absolute normalization (that is to say, the integrated yield of the antineutrino flux) into the analysis; it therefore retains the sensitivity to the mixing angle even at large δm^2 values, where the oscillation structure could no longer be resolved in the energy spectrum and the oscillation limits exclusively depend on the uncertainty in the absolute normalization.

The second approach ("analysis B") uniquely relies on a comparison of measurements taken at different reactor core–detector distances. We are led to limits on the oscillation parameters which are practically independent of the uncertainties in the reactor antineutrino flux and spectrum; other major sources of systematic uncertainties, such as detection efficiencies and the reaction cross section, also are cancelled. The result of this analysis can thus be regarded as free from all these systematic uncertainties and is the safest limit which can be given for the oscillation parameters.

The third approach ("analysis C") is somewhat intermediate between the first two analyses. It uses the shape of the predicted positron spectrum, while leaving the absolute normalization free. The only contribution to the systematic shape uncertainty comes from the precision of the neutrino spectrum extraction method [82].

8.1   Integral test (analysis A)

In a simple two-neutrino oscillation model, the expected positron yield for
the k-th reactor and the j-th energy spectrum bin, can be parametrized as
follows:

$$\overline{X}(E_j, L_k, \theta, \delta m^2) = \tilde{X}(E_j)\overline{P}(E_j, L_k, \theta, \delta m^2), \quad (j = 1, \dots, 7 \quad k = 1, 2)$$

$$(8.1)$$

where $\tilde{X}(E_j)$ is the distance-independent positron yield in the absence of neutrino oscillations defined in the previous Chapter, L_k is the reactor-detector
distance and the last factor represents the survival probability averaged over
the energy bin and the finite detector and reactor core sizes. The procedure to
compute such a probability at varying oscillation parameters is similar to that
used to calculate the burn-up corrections to the positron yields (see §7.6);
the positron spectrum is obtained by Eq.(7.11) after introducing the detector response function r and the size function h; the same procedure applies
to obtain the spectrum for no oscillations; the probability $\overline{P}(E_j, L_k, \theta, \delta m^2)$
results then from the ratio of the j-th bin contents of the two spectra.

In order to test the compatibility of a certain oscillation hypothesis $(\theta, \delta m^2)$
with the measurements, we must build a χ^2 statistic containing 7 experimental yields for each of the two positions L_k (which are listed in Tab. 7.5). For
the sake of simplicity, we grouped these values into a 14-element array X
arranged as follows:

$$\vec{X} = (X_1(E_1), \dots, X_1(E_7), X_2(E_1), \dots, X_2(E_7)), \quad (8.2)$$

and similarly for the associated variances. These components are not independent, as yields corresponding to the same energy bin are extracted simultaneously and the off-diagonal matrix elements σ_{12} (also listed in Tab. 7.5)
are non-vanishing. By combining the statistical variances with the systematic uncertainties related to the neutrino spectrum, the 14×14 covariance
matrix can be written in a compact form as follows:

$$V_{ij} = \delta_{i,j}(\sigma_i^2 + \tilde{\sigma}_i^2) + (\delta_{i,j-7} + \delta_{i,j+7})\sigma_{12}^{(i)} \quad (i, j = 1, \dots, 14), \quad (8.3)$$

where σ_i are the statistical errors associated with the yield array (8.2), $\tilde{\sigma}_i$
are the systematic uncertainties and $\sigma_{12}^{(i)}$ are the covariance of reactor 1 and
2 yield contributions to the i-th energy bin (see Tab. 7.5). These systematic
errors, including the statistical error on the measured β-spectra measured at
ILL [82] as well as the bin-to-bin systematic error inherent in the conversion

procedure, range from 1.4% at $2\,\mathrm{MeV}$ (positron energy) to 7.3% at $6\,\mathrm{MeV}$ and are assumed to be uncorrelated[1].

We still have to take into account the systematic error related to the absolute normalization; combining all the contributions listed in Tab. 8.1, we obtain an overall normalization uncertainty of $\sigma_\alpha = 2.7\%$. We may define the following χ^2 function

$$\chi^2\left(\theta, \delta m^2, \alpha, g\right) =$$
$$\sum_{i=1}^{14}\sum_{j=1}^{14}\left(X_i - \alpha\overline{X}\left(gE_i, L_i, \theta, \delta m^2\right)\right)V_{ij}^{-1}\left(X_j - \alpha\overline{X}\left(gE_j, L_j, \theta, \delta m^2\right)\right) +$$
$$\left(\frac{\alpha - 1}{\sigma_\alpha}\right)^2 + \left(\frac{g - 1}{\sigma_g}\right)^2, \quad (8.4)$$

where α is the absolute normalization constant, g is the energy-scale calibration factor, $L_{i,j} = L_1$ for $i, j \le 7$ and $L_{i,j} = L_2$ for $i, j > 7$. The corresponding uncertainty is $\sigma_g = 1.1\%$, resulting from the accuracy on the energy scale calibration ($16\,\mathrm{KeV}$ at the $2.11\,\mathrm{MeV}$ visible energy line associated with the n-capture on hydrogen) and the 0.8% drift in the Gd-capture line, as measured throughout the acquisition period with high-energy spallation neutrons (see Fig. 7.6). The χ^2 in (8.4) thus contains 14 experimental errors with 2

Table 8.1: Contributions to the overall systematic uncertainty on the absolute normalization factor.

parameter	relative error (%)
reaction cross section	1.9%
number of protons	0.8%
detection efficiency	1.5%
reactor power	0.7%
energy released per fission	0.6%
combined	2.7%

additional parameters, yielding altogether 16 variances. The χ^2 value for a certain parameter set $(\theta, \delta m^2)$ is determined by minimizing (8.4) with respect to the gain factor g and to the normalization α; the minimization then leads to 12 degrees of freedom. The minimum value $\chi^2_{min} = 5.0$, corresponding

---

[1]The extraction of the neutrino spectra from the β measurement at ILL should introduce a slight correlation of the bin-to-bin error systematic error. Nevertheless, the overall uncertainties on the positron yields are dominated by statistical errors, so that neglecting the off-diagonal systematic error matrix does not affect the oscillation test significantly. Also previous reactor experiments, even with much lower statistical errors, did not take this correlation into account.

to a χ^2 probability $P_\chi = 96\%$, is found for the parameters $\sin^2(2\theta) = 0.23$, $\delta m^2 = 8.1 \cdot 10^{-4}\,\mathrm{eV}^2$, $\alpha = 1.012$, $g = 1.006$; the resulting positron yields are shown by solid lines in Fig. 8.1 and superimposed on the data. Also the no-oscillation hypothesis, with $\chi^2(0,0) = 5.5$, $\alpha = 1.008$ and $g = 1.011$, is found to be in excellent agreement with the data ($P_\chi = 93\%$).

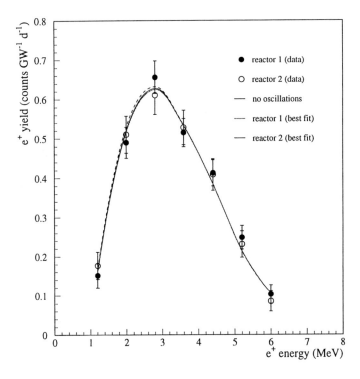

Figure 8.1: Positron yields for reactor 1 and 2; the solid curves represent the predicted positron yields corresponding to the best-fit parameters, the dashed one to the predicted yield for the case of no oscillations.

To test a particular oscillation hypothesis $(\theta, \delta m^2)$ against the parameters of the best fit and to determine the 90% confidence belt, we adopted the Feldman & Cousins prescription [116], now accepted as a standard procedure by the Particle Data Group. The "ordering" principle is based on the logarithm of the ratio of the likelihood functions for the two cases:

$$\lambda(\theta, \delta m^2) = \chi^2(\theta, \delta m^2) - \chi^2_{min} \qquad (8.5)$$

where the mimimum χ^2 value must be searched for within the physical domain $(0 < \sin^2(2\theta) < 1, \delta m^2 > 0)$. Smaller λ values imply a better agreement
of the hypothesis with the data. The λ distribution for the given parameter
set was evaluated by performing a Monte Carlo simulation of a large number
(5000) of experimental positron spectra whose values are scattered around the
predicted positron yields $\overline{X}(E_i, L_i, \theta, \delta m^2)$ with Gaussian-assumed variances
σ_i and correlation coefficients given by (8.3). For each set we extracted the
quantity $\lambda_c(\theta, \delta m^2)$ such that 90% of the simulated experiments have $\lambda < \lambda_c$.
The 90% confidence domain then includes all points in the $(\sin^2(2\theta), \delta m^2)$
plane such that

$$\lambda_{exp}(\theta, \delta m^2) < \lambda_c(\theta, \delta m^2), \tag{8.6}$$

where λ_{exp} is evaluated for the experimental data for each point in the physical domain.

The acceptance domain at the 90% C.L. (solid line) and 95% C.L. are
shown in Fig. 8.2; all parameters lying to the right of the curves are excluded by Chooz with the indicated confidence level, while the parameter
regions on the left are still compatible with our data. The region allowed
by Kamiokande for the $\nu_\mu \rightarrow \nu_e$ oscillations is also shown for comparison;
such a hypothesis, one of the possible explanations for the ν_μ deficit in the
atmospheric neutrino flux, is then definitely excluded. The δm^2 limit at full
mixing is $7 \cdot 10^{-4} \, \text{eV}^2$, to be compared with $9.5 \cdot 10^{-4} \, \text{eV}^2$ previously published [70]; the limit for the mixing angle in the asymptotic range of large
mass differences is $\sin^2(2\theta) = 0.10$, which is twice as low as the previously
published value. Such improvements are essentially due to better statistical
and systematic accuracy.

8.2 Two-distance test (analysis B)

8.2.1 Sensisivity to low δm^2 values

The predicted ratio of the two-reactor positron yields equals the ratio of the
corresponding survival probabilities. At full mixing $(\sin^2(2\theta) = 1)$ and at
low mass differences $(\delta m^2 \approx 10^{-3} \, \text{eV}^2)$, this ratio can be approximated by

$$\overline{R} \approx \left[1 - \left(\frac{1.27\delta m^2 L_1}{E_\nu}\right)^2\right]\left[1 + \left(\frac{1.27\delta m^2 L_2}{E_\nu}\right)^2\right] \approx 1 - 2\left(\frac{1.27\delta m^2}{E_\nu}\right)^2 L\delta L, \tag{8.7}$$

where L is the average reactor-detector distance and δL is the difference
of the two distar ces. Therefore an experiment, whose measured ratio is

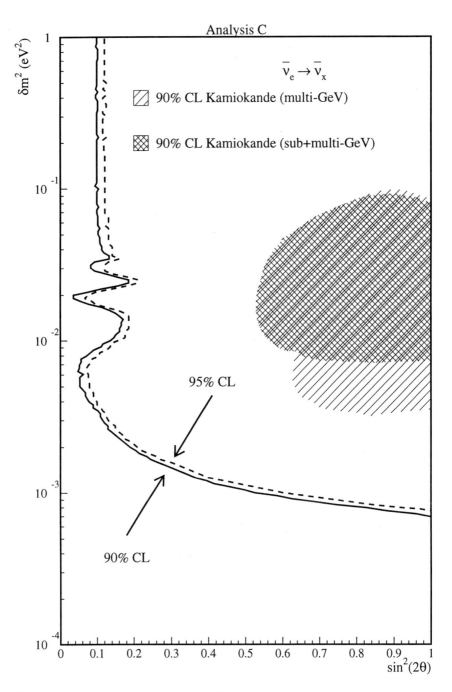

Figure 8.2: Exclusion plot for the oscillation parameters based on the absolute comparison of measured vs. expected positron yields.

affected by the uncertainty σ, retains its sensitivity to oscillations down to mass-difference values as low as

$$\delta m^2 \approx \frac{E_\nu}{1.27}\sqrt{\frac{k\sigma}{2L\delta L}}, \tag{8.8}$$

k being the number of standard deviations corresponding to the chosen confidence level[2]. This value can be connected to the sensitivity limit δm_0^2 inherent in analysis A by the relation

$$\delta m^2 \approx \sqrt{\frac{L}{2\delta L}}\delta m_0^2 \approx 2\delta m_0^2 \approx 1.5\cdot10^{-3}\,\mathrm{eV^2} \tag{8.9}$$

Although twice as large, this limit is lower than the lowest δm^2 value allowed by Kamiokande (see Fig. 8.2).

8.2.2 Ratios of energy spectra

The ratio $R(E_i) \equiv X_1(E_i)/X_2(E_i)$ of the measured positron yields must be compared with the expected values; since the expected yields are common to both reactors in the case of no-oscillations, the expected ratio reduces to the ratio of the average survival probabilities at the i-th energy bin. We can then build the following χ^2 statistic:

$$\chi^2 = \sum_{i=1}^{7}\left(\frac{R(E_i) - \overline{R}(E_i,\theta,\delta m^2)}{\delta R(E_i)}\right)^2 \tag{8.10}$$

where $\delta R(E_i)$ is the statistical uncertainty on the measured ratio. The minimum χ^2 value ($\chi^2_{min} = 0.78$ over 5 d.o.f.) occurs at $\sin^2(2\theta) = 1$ and $\delta m^2 = 0.6$; the agreement of the no-oscillation hypothesis is still excellent (see Fig. 8.3), as $\chi^2(0,0) = 1.29$.

We adopted the same procedure described in the previous Section to determine the confidence domain in the $(\sin^2(2\theta),\delta m^2)$ plane; for each point in this plane we simulated by the Monte Carlo method the results of 5000 experiments. If the positron yields of both reactors are gaussian distributed around the predicted values $\overline{X}_{1,2}$, and if $\overline{X}_2/\sigma_2$ is sufficiently large so that X_2 has practically only positive values, then the variable

$$Z = \frac{\dfrac{\overline{X}_1}{X_1} - \dfrac{\overline{X}_2}{X_2}}{\sqrt{\dfrac{\sigma_1^2}{X_1^2} + \dfrac{\sigma_2^2}{X_2^2} - 2\dfrac{\sigma_{12}}{X_1X_2}}} = \frac{\overline{X}_1 - \overline{X}_2R}{\sqrt{\sigma_1^2 + \sigma_2^2R^2 - 2\sigma_{12}R}} \tag{8.11}$$

---

[2]For instance $k = 1.64$, if we assume that the measured ratio is Gaussian-distributed around the predicted value.

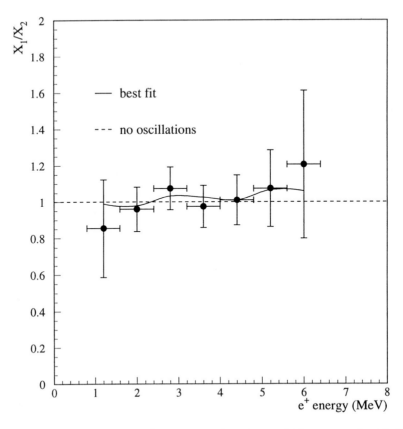

Figure 8.3: Measured ratio of experimental positron yield, compared with the predicted ratio in the best oscillation hypothesis (solid line) and in the case of no oscillations (dashed line).

is normally distributed with zero mean and unit variance [110]. So we can use a normal random generator to extract Z and invert (8.11) to determine the ratio $R(E_i)$.

The exclusion plot so-obtained is shown in Fig. 8.4; the contour lines of the 90% and 95% C.L. are drawn. Although less powerful than the previous analysis, the region excluded by this oscillation test almost completely covers the one allowed by Kamiokande.

8.2.3   Comparison of the spectra

In building the χ^2 statistic (8.10) we assumed that the ratios $R(E_i)$ were Gaussian distributed around the predicted values, which is true with good approximation as long as the statistical error over the positron yield of either

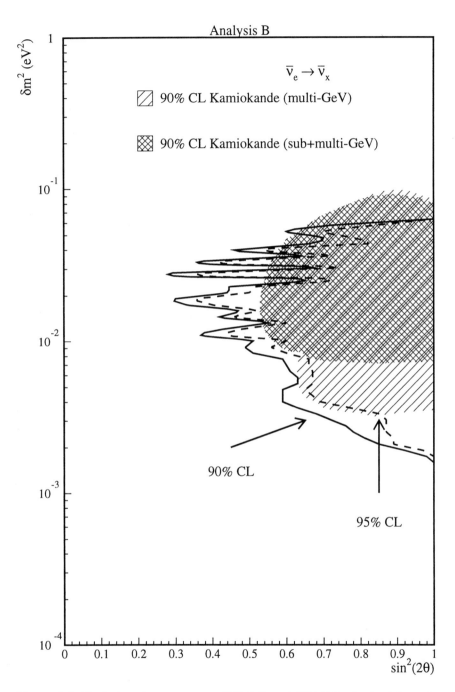

Figure 8.4: Exclusion plot contours at 90% C.L. and 95% C.L. obtained from the ratios of the positron yields induced from the two reactors.

reactor is lower than 10%; this condition is not fulfilled by the lower and upper energy bins (where the statistical uncertainty is $\approx 20\%$). As a check of previous results, we then decided to perform an additional two-reactor comparison test where the ratios of the positron yields were no longer used. By using the same notation as in Eqs.(8.4,8.1), we built the following χ^2 statistic

$$\chi^2(\theta, \delta m^2, \vec{\tilde{X}}) =$$
$$\sum_{i=1}^{14}\sum_{j=1}^{14} \left(\frac{X_i - \tilde{X}_i \overline{P}(E_i, L_i, \theta, \delta m^2)}{\sigma_i} \right) \rho_{ij}^{-1} \left(\frac{X_j - \tilde{X}_j \overline{P}(E_j, L_j, \theta, \delta m^2)}{\sigma_j} \right),$$
$$\tag{8.12}$$

where σ_i included only the statistical uncertainty on the yield X_i and the no-oscillation yields $\tilde{X}_i$ were considered as free parameters to be determined by the fit.

This "comparison" test gave results very similar to that obtained by the ratio test, as shown in Fig. 8.5; the most significant differences arise in the region of low δm^2, where the oscillation amplitude is maximal at the first energy bin.

8.3   Shape test (analysis C)

The shape test is similar to analysis A, the only difference being related to the hypothesis on the absolute normalization. In the analysis A we fixed the integral counting rate to be distributed around the predicted value ($\alpha = 1$), with $\sigma_\alpha = 2.7\%$ systematic uncertainty; in the shape test, on the contrary, we were allowed to give up any constraint on the normalization parameter (which is equivalent to having $\sigma_\alpha = \infty$). The χ^2 statistic for this test then follows from (8.4) after erasing the term depending on the normalization; so we can write

$$\chi^2(\theta, \delta m^2, \alpha, g) =$$
$$\sum_{i=1}^{14}\sum_{j=1}^{14} \left(X_i - \alpha \overline{X}\left(g E_i, L_i, \theta, \delta m^2\right) \right) V_{ij}^{-1} \left(X_j - \alpha \overline{X}\left(g E_j, L_j, \theta, \delta m^2\right) \right) +$$
$$\left(\frac{g-1}{\sigma_g} \right)^2, \quad \tag{8.13}$$

The (8.13) has a minimum value $\chi^2_{min} = 2.64$ (over 11 degrees of freedom) at $\sin^2(2\theta) = 0.23$, $\delta m^2 = 2.4 \cdot 10^{-2}\,\text{eV}^2$ and $g = 1.008$; the null hypothesis gives instead $\chi^2(0,0) = 5.5$ with $g = 1.006$. The exclusion plot, obtained

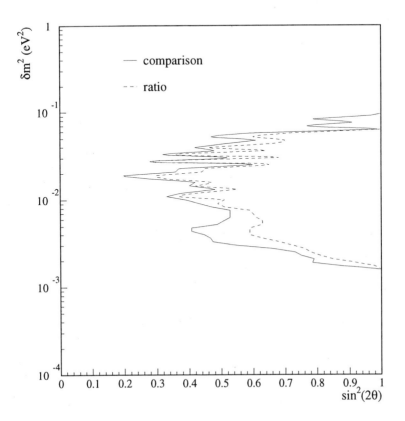

Figure 8.5: Exclusion plots at 90% C.L. for the two-distance tests, by using the comparison (solid line) and the ratio (dashed line) of the two reactor spectra.

according to the Feldman–Cousins prescriptions, is shown in Fig. 8.6 and compared to the results of the other tests. As in the case of the two-distance test, its sensitivity to oscillations turns off at large squared mass-difference values ($\delta m^2 \gtrsim 0.1\,\mathrm{eV^2}$), where the oscillation length

$$L_{osc}(\mathrm{m}) = \frac{2.48 E_\nu(\mathrm{MeV})}{\delta m^2(\mathrm{eV^2})} \lesssim 100\,\mathrm{m} \tag{8.14}$$

becomes much lower than the average reactor–detector distance. The $\sin^2(2\theta)$ limit at the maximum oscillation probability is similar to that obtained by the integral test.

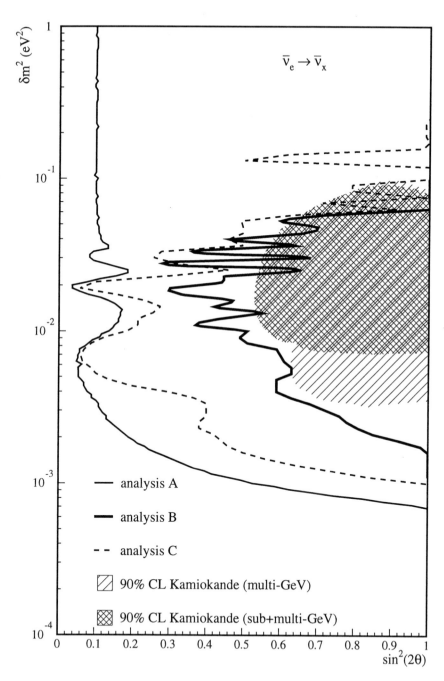

Figure 8.6: Exclusion plot contours at 90% C.L. obtained by the three analyses presented.

8.4 Implications of the Chooz results

Before the first Chooz results were published, the atmospheric neutrino anomaly could be explained also in terms of $\nu_\mu \rightarrow \nu_e$ oscillations. The importance of these results on neutrino oscillations has been pointed out by several authors [117, 118, 22, 53]. In a three-flavour neutrino mixing frame, the see-saw mechanism indicates a mass hierarchy ($m_1 \ll m_2 \ll m_3$) from which

$$\delta m_{12}^2 \ll \delta m_{23}^2. \tag{8.15}$$

The only available possibility for the explanation of solar and atmospheric neutrino anomalies through neutrino oscillations is that δm_{12} is relevant for the transitions of solar neutrinos and δm_{13} is that probed by atmospheric and LBL neutrino experiments. Under the approximation (8.15) it is possible to show that the CP-violating phase in the 3×3 mixing matrix does not give rise to observable effects, and that the mixing angle θ_{12} (associated with the lower mass states) can be rotated away in the atmospheric neutrino analysis. Therefore, condition (8.15) implies that the transition probability in atmospheric and LBL experiments depends only on the largest mass squared difference δm_{23} and the elements $U_{\alpha 3}$ connecting flavour neutrinos with ν_3; it assumes then the simple form

$$P_{\nu_\alpha \rightarrow \nu_\beta} = \begin{cases} 1 - 4U_{\alpha 3}^2(1 - U_{\alpha 3}^2)\sin^2\left(\dfrac{1.27\delta m^2(\text{eV}^2)L(\text{m})}{E(\text{MeV})}\right) & \text{if } \alpha = \beta, \\[3mm] 4U_{\alpha 3}^2 U_{\beta 3}^2 \sin^2\left(\dfrac{1.27\delta m^2(\text{eV}^2)L(\text{m})}{E(\text{MeV})}\right) & \text{if } \alpha \neq \beta. \end{cases} \tag{8.16}$$

In particular, the survival probability reduces to the usual two-flavour formula (2.8) with $\sin^2 2\theta = 4U_{\alpha 3}^2(1 - U_{\alpha 3}^2)$. Therefore, information on the parameter U_{e3} can be obtained from the Chooz exclusion plot. The upper limit for $\sin^2 2\theta$ implied by the exclusion plot in Fig. 8.2 is $\sin^2 2\theta \lesssim 0.1$ for $\delta m_{13} \gtrsim 2 \cdot 10^{-3}$; it follows that

$$U_{e3}^2 < 0.03 \quad or \quad U_{e3}^2 > 0.97 \tag{8.17}$$

Large values of values of U_{e3}, those allowed by the second inequality in eq.(8.17), are in fact excluded by the solar neutrino data [53]; the solar ν_e survival probability would be larger than 0.95, which is incompatible with the deficit observed in all solar neutrino experiments. The Chooz result then constrains the mixing of electron neutrinos with other flavours, in the atmospheric and LBL range, to small values.

It has been remarked [118] that Super-Kamiokande data alone do not exclude sizable ν_e mixing. This can be seen in Fig. 8.7, where the confidence regions for separate and combined Super-Kamiokande and Chooz data are shown for different values of δm_{23} (the mass difference relevant for atmospheric neutrinos). Each set of mixing parameters $(U_{e3}, U_{\mu3}, U_{\tau3})$ is associated with a point embedded in an triangle graph (equilateral due to the unitarity condition), whose corners represent the flavour eigenstates; by convention ν_e is assigned the upper corner. The distance from the three sides are equal to $U_{e3}^2, U_{\mu3}^2, U_{\tau3}^2$; in a two-flavour scheme ($U_{\alpha3} = 0$), the point is bound on the side connecting the two mixed flavour states (and the mean point of that side is associated with the maximum mixing hypothesis).

The Chooz result excludes large horizontal stripes in the triangle plot, according to eq.(8.16); the stripe becomes narrower and narrower at lower mass δm_{23} values, as a consequence of the reduced sensitivity to $\sin^2 2\theta$. The Super-Kamiokande allowed region lies on the triangle base and protrudes towards the centre, which implies a non-negligible $\nu_\mu \rightarrow \nu_e$ oscillation problability. In the combined analysis graph instead, the allowed region is significantly flattened on the base, thus indicating a dominance of the $\nu_\mu \rightarrow \nu_\tau$ maximum mixing hypothesis.

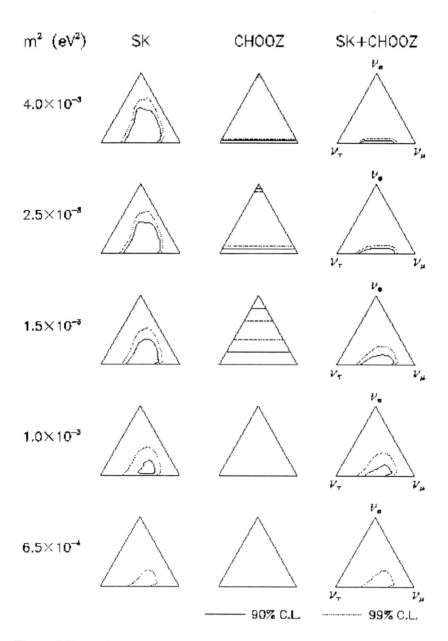

Figure 8.7: Results of a three flavour mixing analysis of separate and combined Super-Kamiokande and Chooz data, for five representative values of δm_{23}. The analysis concerns the 33 kTy data sample for Super-Kamiokande and the first Chooz result.

Chapter 9

Conclusions

The Chooz experiment stopped taking data in July 1998, about 5 years (which is a rare case indeed nowadays!) after the submission of the proposal to approval. With more than 1-year data taking, the statistical error (2.8%) on the neutrino flux fulfilled the goal (3%) of the proposal (2 ÷ 4%) [5]; an accurate estimates of the detection efficiencies as well as a precise measurements of the detector parameters allowed us to keep also the systematic uncertainty (2.7%) below expectations (3.2%).

We found (at 90% confidence level) no evidence for neutrino oscillations in the $\bar{\nu}_e$ disappearance mode, for the parameter region given by approximately $\delta m^2 > 7 \cdot 10^{-4}\,\text{eV}^2$ for maximum mixing, and $\sin^2 2\theta = 0.10$ for large δm^2. Lower sensitivity results, based only on the comparison of the positron spectra from the two different-distance nuclear reactors (and therefore independent of the absolute normalization of the $\bar{\nu}_e$ flux, the number of protons and the detector efficiencies) were also presented. All these results have just been submitted to publication [119].

The impact of Chooz for atmospheric neutrino oscillation searches has been recognized by many authors ([117, 118, 53, 22] to mention just a few). Our result excludes that the atmospheric neutrino anomaly can be explained in terms of $\nu_\mu \rightarrow \nu_e$ oscillations, thus leaving, in a three-flavour mixing scheme, the $\nu_\mu \rightarrow \nu_\tau$ possibility.

Many cross-checks were performed on the data to test the internal consistence of our results and improve the reliability of our results. As a byproduct, we have shown that the use of reaction (2.22) allowed us to locate the $\bar{\nu}_e$ source within a cone of half-aperture $\simeq 18°$ at 68% confidence level. We have also shown that this technique (essentially based on the neutron boost along the incoming $\bar{\nu}_e$ direction) provides a powerful tool to large scintillator detectors for pointing towards neutrino astrophysical sources, such as Supernovæ. The Chooz measurement as well as its implication to Supernova localization has been recently submitted to publication [120].

Acknowledgements

At the end of this work I feel bound to thank all people that supported my efforts all along these years.

I feel extremely lucky in experiencing the guidance of my advisor, Carlo Bemporad. Not many colleagues of mine can be proud of being guided with as much care as I have been. I am indebted to him for teaching me much physics and an entire philosophy of scientific thinking. I would have never arrived at this goal without his continuous support and encouragement. Still I have to thank him for giving me the opportunity of working in a research team, remarkable not only on a scientific ground (which is widely recognized), but also for its humanity. Roberto Pazzi, Alessandro Baldini, Marco Grassi and Fabrizio Cei helped me in solving so many problems that I cannot list; in short, all that contained in my thesis (although not explicitly said before) is a result of everyone's effort. I have to extend my thanks to the Trieste group, and in particular to his leader, Gianrossano Giannini, for the fruitful co-operation and stimulating discussions. I have also benefitted from the work of my younger colleagues Giovanni Pieri and Luca Foresti, and also Sandra Parlati, Stefano Stalio and Emanuela Pitzalis, who collaborated with the Chooz–Pisa group during these years and contributed to the success of the experiment.

For the same reason I am grateful to all the members of the Chooz collaboration. Since I cannot cite them individually, I will mention the french spokesman, Yves Déclais. Without his dedication and his realization capabilities the experiment could not have been carried out.

It is a pleasure to remember here all my friends in Biccari, including the just-arrived Pietro and Andrea. I would like to ideally join my fraternal friend Ernesto, with whom I shared joys and sorrows during the years at the University. Although the chances to meet are less and less frequent, my affection on them is unchanged. It is for that reason that my bond "to the place where I belonged" will never loosen.

I owe much to all my family, who always share the ups and downs of my life. I will never repay my parents enough for what they have done for me and my brothers. This thesis is dedicated to them for their love and patience.

Bibliography

[1] F. Reines and C. L. Cowan, Phys. Rev. **90** (1953),492

[2] Y. Fukuda *et al.*, Phys. Lett. **B335** (1994),237

[3] R. Becker-Szendy *et al.*, Phys .Rev. **D46** (1992)3720

[4] W. W. M. Allison *et al.*, Phys. Lett. **B391** (1997),491

[5] The Chooz experiment proposal, C. Bemporad, Y. Declais and R. Steinberg co-spokesmen (1993)

[6] G. Zacek *et al.*, Phys. Rev. **D34** (1986),2621

[7] B. Achkar *et al.*, Nucl. Phys. **B434** (1995),503

[8] V. Lobashev, Proc. of the 17th International Workshop on Weak Interactions and Neutrinos WIN99, Capetown 1999, to be published

[9] K. Assamagan *et al.*, Phys. Rev. **D53** (1996),6065

[10] R. Barate *et al.*, Europ. Phys. J. **C2** (1998),395

[11] S. L. Glashow, Nucl. Phys. **B22** (1961),579
A. Salam and J. C. Ward, Phys. Lett. **13** (1964),168
S. Weinberg, Phys. Rev. Lett. **19** (1967),1264

[12] S. L. Glashow, J. Iliopoulos and L. Maiani, Phys. Rev. **D2** (1970),1285

[13] E. S. Abers and B. W. Lee, Phys. Rep. **C9** (1973),1

[14] L. B. Okun, *Leptoni e quark*, Editori Riuniti Roma (1986)

[15] P. W. Higgs, Phys. Rev. Lett. **12** (1964),132
F. Englert and R. Brout, Phys. Rev. Lett. **13** (1964),321

[16] N. Cabibbo, Phys. Rev. Lett. **10** (1963),531
M. Kobayashi and K. Maskawa, Prog. Theo. Phys. **49** (1973),652

[17] F. Boehm and P. Vogel, *Physics of massive neutrinos*, Cambridge University Press, Cambridge (1992)

[18] K. S. Babu, B. Dutta and R. N. Mohapatra, preprint hep-ph/9904366 (1999)

[19] R. N. Mohapatra and P. B. Pal, *Massive neutrinos in Physics and Astrophysics*, World Scientific Singapore (1991)

[20] E. W. Kolb and M. Turner, *The Early Universe*, Addison-Wesley, Redwood City, California (1990)

[21] S. Weinberg, *Gravitation and Cosmology*, Wiley, New York (1972)

[22] R. D. Peccei, preprint hep-ph/9906509 (1999)

[23] W. L. Freedman, preprint astro-ph/9905222 (1999)

[24] B. Pontecorvo, Sov. Phys. JETP **6** (1958),429

[25] Z. Maki, M. Nakagawa and S. Sakata, Prog. Theor. Phys. **28** (1962),870

[26] B. Pontecorvo, Sov. Phys. JETP **26** (1968),984

[27] L. Wolfenstein, Phys. Rev. **D17** (1978),2369

[28] S. P. Mikheyev and A. Yu. Smirnov, Sov. J. Nucl. Phys. **42** (1985)913

[29] T. K. Gaisser, *Cosmic Rays and Particle Physics*, Cambridge University Press, Cambridge (1990)

[30] Y. Fukuda *et al.*, Phys. Rev. Lett. **81** (1998),1562

[31] M. A. Aglietta *et al.*, Phys. Lett. **B280** (1992),146

[32] K. Daum *et al.*, Zeit. Phys. **C66** (1995),417

[33] T. J. Haines *et al.*, Phys. Rev. Lett. **57** (1986),1986

[34] K. S. Hirata *et al.*, Phys. Lett. **B280** (1992),146

[35] M. Ambrosio *et al.*, Phys. Lett. **B434** (1998),451

[36] J. N. Bahcall, *Neutrino Astrophysics*, Cambridge University Press, Cambridge (1989)

[37] R. Davis, D. S. Harmer and K. C. Hoffman, Phys. Rev. Lett. **20** (1968),1205

[38] J. N. Bahcall, S. Basu and M. H. Pinsonneault, Phys. Lett. **B433** (1998),1

[39] B. T. Cleveland *et al.*, Astrophys. J. **496** (1998),505

[40] W. Hampel *et al.*, Phys. Lett. **B388** (1996),384

[41] D. N. Abdurashitov *et al.*, Phys. Rev. Lett. **77** (1996),4708

[42] K. S. Hirata *et al.*, Phys. Rev. Lett. **77** (1996),1683

[43] Y. Suzuki, Proc. of the 18th Neutrino Conference, Takayama 1998, to be published

[44] S. Turck-Chièze *et al.*, Phys. Rep. **230** (1993),57

[45] B. Ricci *et al.*, Phys. Lett. **B407** (1997),155

[46] V. Castellani *et al.*, Phys. Rep. **281** (1997),309

[47] J. N. Bahcall, P. I. Krastev and A. Yu. Smirnov, Phys. Rev. **D58** (1998),096016

[48] A. McDonald, Proc. of the 18th Neutrino Conference, Takayama 1998, to be published

[49] L. Oberauer, Proc. of the 18th Neutrino Conference, Takayama 1998, to be published

[50] C. Athanassopoulos *et al.*, Phys. Rev. **C54** (1996),2685

[51] C. Athanassopoulos *et al.*, Phys. Rev. **C58** (1998),2489

[52] B. Zeitnitz, Proc. of the 18th Neutrino Conference, Takayama 1998, to be published

[53] S. M. Bilenky, C. Giunti and W. Grimus, preprint hep-ph/9809501 (1998)

[54] BooNE proposal, available at http://nu1.lampf.lanl.gov/BooNE

[55] I-216 proposal, available at http://chorus01.cern.ch/ pzucchel/loi

[56] NESS proposal, available at http://www.isis.rl.ac.uk/ess/neut

[57] The CHORUS collaboration, preprint hep-ph/9907015 (1999)

[58] J. Altegoer *et al.*, Phys. Lett. **B431** (1998),219

[59] Y. Suzuki, Proc. of the 17th Neutrino Conference, Helsinki 1996, World Scientific (1997)

[60] D. Ayres et al., report NUMI-L-63 (1995)

[61] H. Shibuya et al., report CERN-SPSC-97-24 (1997)

[62] G. Barbarino et al., report INFN/AE-96/11 (1996)

[63] T. Ypsilantis et al., report CERN-LAA/96-13 (1996)

[64] A. Baldini et al., report LNGS-LOI 98/13 (1998)

[65] C. Rubbia, Nucl. Phys. Proc. Suppl. **48** (1996),172

[66] Z. D. Greenwood et al., Phys. Rev. **D53** (1996),6054

[67] A. I. Afonin et al., Sov. Phys. JETP **66** (1988),213

[68] G. S. Vidyakin et al., JETP Lett. **59** (1994),390

[69] Y. Wang, Proc. of the XXXIV Rencontres de Moriond on Electroweak Interactions and Unified Theories, 1999, to be published

[70] M. Apollonio et al., Phys. Lett. **B420** (1998),397

[71] A. Suzuki, Proc. of the 18th Conference on Neutrino Telescopes, Venice 1999, to be published

[72] B. Pontecorvo, Nat. Res. Council Canada Rep. (1946),205; Helv. Phys. Acta Suppl. vol. 3 (1950),97

[73] A. A. Kuvshinnikov et al., Sov. J. Nucl. Phys. **52** (1990),300

[74] Dossier de presse E.D.F., *Mise en service de Chooz B1* (1996)

[75] M. F. James, J. Nucl. Energy **23** (1969),517

[76] R. W. King and J. F. Perkins, Phys. Rev. **112** (1958),963

[77] W. I. Kopeikin et al., Sov. J. Nucl. Phys. **32**(1980),2810

[78] F. T. Avignone III and Z. D. Greenwood Phys. Rev. **C22** (1980),594

[79] P. Vogel et al., Phys. Rev. **C24** (1981),1543

[80] H. V. Klapdor and J. Metzinger, Phys. Rev. Lett. **48** (1982),127

[81] O. Tengblad et al., Nucl. Phys. **A503** (1989),136

[82] F. von Feilitzsch *et al.*, Phys. Lett. **B118** (1982),162

[83] R. E. Carter *et al.*, Phys. Rev. **113** (1959),280

[84] B. R. Davis *et al.*, Phys. Rev. **C19** (1979),2259

[85] A. A. Borovoi *et al.*, Sov. J. Nucl. Phys. **37** (1983),801

[86] K. Schreckenbach *et al.*, Phys. Lett. **B160** (1985),325

[87] A. A. Hahn *et al.*, Phys. Lett. **B218** (1989),365

[88] B. Achkar *et al.*, Phys. Lett. **B374** (1996),243

[89] Y. Declais *et al.*, Phys. Lett. **B338** (1994),383

[90] V. I. Kopeikin *et al.*, Kurchatov Internal Report IAE-6026/2 (1997)

[91] D. H. Wilkinson, Nucl. Phys. **A377** (1982),474

[92] Particle Data Group, Europ. Phys. J. **C3** (1998),50

[93] P. Vogel, Phys. Rev. **D29** (1984),1918

[94] S. A. Fayans, Sov. J. Nucl. Phys. **42** (1985),540

[95] F. Reines *et al.*, Phys. Rev. Lett. **45** (1980),1307

[96] Y. Declais, E. Duverney and A. Oriboni, Chooz Design Technical Report (1995)

[97] Laboratory of Organic Chemistry, ETH Zürich (kindness of prof. A. Pretsch)

[98] *Service central d' analyse, laboratoire du CNRS*, Vernaison (France)

[99] R. S. Raghavan, Phys. Rev. Lett. **78** (1997),3618

[100] A. Baldini *et al.*, Nucl. Instr. Meth. **A372** (1996),207

[101] The Thorn EMI Inc. photomultipliers and accessories catalogue

[102] D. Nicolò, Pretesi di Perfezionamento, Scuola Normale Superiore (1995)

[103] GEANT Reference Manual vers. 3.21 (1993)

[104] R. Liu, MACRO internal memo 11/93 (1993)

[105] H. de Kerret, B. Lefièvre, LPC internal report 88-01

[106] S. F. Mughabghab, *Neutron cross sections*, ed. Academic Press

[107] L. V. Groshev *et al.*, Nucl. Data Tab. **A5** (1968),1

[108] R. R. Spencer, R. Gwin and R. Ingle, Nucl. Sci. and Eng. **80** (1982),603

[109] CNAPS manual, Adaptive Solution

[110] W. T. Eadie *et al.*, *Statistical methods of experimental physics*, North-Holland, Amsterdam (1971)

[111] S. Baker and R. Cousins, Nucl. Instr. Meth. **221** (1984),437

[112] F. James, MINUIT Reference Manual, vers. 94.1 (1994)

[113] P. Vogel and J. F. Beacom, preprint hep-ph/9903554 (1999), to appear on Phys. Rev. D

[114] E. Amaldi, *The production and slowing down of neutrons*, S.Flügge, Encyclopedia of Physics (Springer–Verlag ed.), vol. **38.2** (1959)

[115] S. A. Bludman and P. J. Schinder, Astrophys. J. **326** (1988),265

[116] G. J. Feldman and R. D. Cousins, Phys. Rev. **D57** (1998),3873

[117] R. Barbieri *et al.*, preprint hep-ph 9807235 (1998)

[118] G. L. Fogli *et al.*, preprint hep-ph/9808205 (1998)

[119] M. Apollonio *et al.*, preprint hep-ph/9907037 (1999), submitted to Phys. Lett. B

[120] M. Apollonio *et al.*, preprint hep-ph/9906011 (1999), submitted to Phys. Rev. D

Elenco delle Tesi di perfezionamento della Classe di Scienze
pubblicate dall'Anno Accademico 1992/93

HISAO FUJITA YASHIMA, *Equations de Navier-Stokes stochastiques non homogènes et applications*, 1992.

GIORGIO GAMBERINI, *The minimal supersymmetric standard model and its phenomenological implications*, 1993.

CHIARA DE FABRITIIS, *Actions of Holomorphic Maps on Spaces of Holomorphic Functions*, 1994.

CARLO PETRONIO, *Standard Spines and 3-Manifolds*, 1995.

MARCO MANETTI, *Degenerations of Algebraic Surfaces and Applications to Moduli Problems*, 1995.

ILARIA DAMIANI, *Untwisted Affine Quantum Algebras: the Highest Coefficient of* det H_η *and the Center at Odd Roots of 1*, 1995.

FABRIZIO CEI, *Search for Neutrinos from Stellar Gravitational Collapse with the MACRO Experiment at Gran Sasso*, 1995.

ALEXANDRE SHLAPUNOV, *Green's Integrals and Their Applications to Elliptic Systems*, 1996.

ROBERTO TAURASO, *Periodic Points for Expanding Maps and for Their Extensions*, 1996.

YURI BOZZI, *A study on the activity-dependent expression of neurotrophic factors in the rat visual system*, 1997.

MARIA LUISA CHIOFALO, *Screening effects in bipolaron theory and high-temperature superconductivity*, 1997.

DOMENICO M. CARLUCCI, *On Spin Glass Theory Beyond Mean Field*, 1998.

RENATA SCOGNAMILLO, *Principal G-bundles and abelian varieties: the Hitchin system*, 1998.

GIACOMO LENZI, *The MU-calculus and the Hierarchy Problem*, 1998.

GIORGIO ASCOLI, *Biochemical and spectroscopic characterization of CP20, a protein involved in synaptic plasticity mechanism*, 1998.

FABIO PISTOLESI, *Evolution from BCS Superconductivity to Bose-Einstein Condensation and Infrared Behavior of the Bosonic Limit*, 1998.

LUIGI PILO, *Chern-Simons Field Theory and Invariants of 3-Manifolds*, 1999.

PAOLO ASCHIERI, *On the Geometry of Inhomogeneous Quantum Groups*, 1999.

SERGIO CONTI, *Ground state properties and excitation spectrum of correlated electron systems*, 1999.

GIOVANNI GAIFFI, *De Concini-Procesi models of arrangements and symmetric group actions*, 1999.

DONATO NICOLÒ, *Search for neutrino oscillations in a long baseline experiment at the Chooz nuclear reactors*, 1999.

"CompoMat" Loc. Braccone, 02040 Configni (RI), Italy
Finito di stampare per conto della "CompoMat" dalla Nuova Grafica 86 nel febbraio 2000